LONDON MATHEMATICAL SOCIETY LECTURE NOTE SERIES

Managing Editor: Professor I.M.James,
Mathematical Institute, 24-29 St Giles, Oxford



1. General cohomology theory and K-theory, P.HILTON
4. Algebraic topology: a student's guide, J.F.ADAMS
5. Commutative algebra, J.T.KNIGHT
8. Integration and harmonic analysis on compact groups, R.E.EDWARDS
9. Elliptic functions and elliptic curves, P.DU VAL
10. Numerical ranges II, F.F.BONSALL & J.DUNCAN
11. New developments in topology, G.SEGAL (ed.)
12. Symposium on complex analysis, Canterbury, 1973, J.CLUNIE & W.K.HAYMAN (eds.)
13. Combinatorics: Proceedings of the British combinatorial conference 1973, T.P.McDONOUGH & V.C.MAVRON (eds.)
14. Analytic theory of abelian varieties, H.P.F.SWINNERTON-DYER
15. An introduction to topological groups, P.J.HIGGINS
16. Topics in finite groups, T.M.GAGEN
17. Differentiable germs and catastrophes, Th.BROCKER & L.LANDER
18. A geometric approach to homology theory, S.BUONCRISTIANO, C.P.ROURKE & B.J.SANDERSON
20. Sheaf theory, B.R.TENNISON
21. Automatic continuity of linear operators, A.M.SINCLAIR
23. Parallelisms of complete designs, P.J.CAMERON
24. The topology of Stiefel manifolds, I.M.JAMES
25. Lie groups and compact groups, J.F.PRICE
26. Transformation groups: Proceedings of the conference in the University of Newcastle upon Tyne, August 1976, C.KOSNIOWSKI
27. Skew field constructions, P.M.COHN
28. Brownian motion, Hardy spaces and bounded mean oscillation, K.E.PETERSEN
29. Pontryagin duality and the structure of locally compact abelian groups, S.A.MORRIS
30. Interaction models, N.L.BIGGS
31. Continuous crossed products and type III von Neumann algebras, A.VAN DAELE
32. Uniform algebras and Jensen measures, T.W.GAMELIN
33. Permutation groups and combinatorial structures, N.L.BIGGS & A.T.WHITE
34. Representation theory of Lie groups, M.F.ATIYAH et al.
35. Trace ideals and their applications, B.SIMON
36. Homological group theory, C.T.C.WALL (ed.)
37. Partially ordered rings and semi-algebraic geometry, G.W.BRUMFIEL
38. Surveys in combinatorics, B.BOLLOBAS (ed.)
39. Affine sets and affine groups, D.G.NORTHCOTT
40. Introduction to $H_p$ spaces, P.J.KOOSIS
41. Theory and applications of Hopf bifurcation, B.D.HASSARD, N.D.KAZARINOFF & Y-H.WAN
42. Topics in the theory of group presentations, D.L.JOHNSON
43. Graphs, codes and designs, P.J.CAMERON & J.H.VAN LINT
44. Z/2-homotopy theory, M.C.CRABB
45. Recursion theory: its generalisations and applications, F.R.DRAKE & S.S.WAINER (eds.)
46. p-adic analysis: a short course on recent work, N.KOBLITZ
47. Coding the Universe, A. BELLER, R. JENSEN & P. WELCH
48. Low-dimensional topology, R. BROWN & T.L. THICKSTUN (eds.)
49. Finite geometries and designs, P. CAMERON, J.W.P. HIRSCHFELD & D.R. HUGHES (eds.)
50. Commutator Calculus and groups of homotopy classes, H.J. BAUES
51. Synthetic differential geometry, A. KOCK
52. Combinatorics, H.N.V. TEMPERLEY (ed.)
53. Singularity theory, V.I. ARNOLD
54. Markov processes and related problems of analysis, E.B. DYNKIN
55. Ordered permutation groups, A.M.W. GLASS
56. Journees arithmetiques 1980, J.V. ARMITAGE (ed.)
57. Techniques of geometric topology, R.A. FENN
58. Singularities of differentiable functions, J. MARTINET
59. Applicable differential geometry, F.A.E. PIRANI and M. CRAMPIN
60. Integrable systems, S.P. NOVIKOV et al.

61. The core model, A. DODD
62. Economics for mathematicians, J.W.S. CASSELS
63. Continuous semigroups in Banach algebras, A.M. SINCLAIR
64. Basic concepts of enriched category theory, G.M. KELLY
65. Several complex variables and complex manifolds I, M.J. FIELD
66. Several complex variables and complex manifolds II, M.J. FIELD
67. Classification problems in ergodic theory, W. PARRY & S. TUNCEL

London Mathematical Society Lecture Note Series. 48

# Low-Dimensional Topology

## Volume 1 of the Proceedings of the Conference on Topology in Low Dimension, Bangor, 1979

Edited by
R. BROWN and T.L. THICKSTUN
School of Mathematics and Computer Science
University College of North Wales
Bangor

CAMBRIDGE UNIVERSITY PRESS

CAMBRIDGE

LONDON NEW YORK NEW ROCHELLE

MELBOURNE SYDNEY

CAMBRIDGE UNIVERSITY PRESS
Cambridge, New York, Melbourne, Madrid, Cape Town, Singapore, São Paulo, Delhi

Cambridge University Press
The Edinburgh Building, Cambridge CB2 8RU, UK

Published in the United States of America by Cambridge University Press, New York

www.cambridge.org
Information on this title: www.cambridge.org/9780521281461



First published 1982
Re-issued in this digitally printed version 2009

*A catalogue record for this publication is available from the British Library*

*Library of Congress Catalogue Card Number: 81-2664*

ISBN 978-0-521-28146-1 paperback

# Contents

## Dedication

The Conference was dedicated to Peter Stefan, Lecturer at the University College of North Wales, who fell to his death in June 1978, while climbing by himself on Tryfan. He was known to most of the participants from Britain, and to many from France owing to his leave of absence at the Institut des Hautes Études, Bures-sur-Yvette, in 1976/7. He had been a prime mover in obtaining (with R. Brown) support for Tom Thickstun for research in 3-manifolds at Bangor, and the idea for this conference had been agreed between us before his death.

A survey of his work is given in the Obituary notice in the Bulletin London Math. Soc. 13 (1981) 170-172.

## Acknowledgements

We are grateful to:

The London Mathematical Society, who accepted fully our request for partial financial support of the Conference, and under whose auspices this book is appearing.

Dr. G.P. Scott, who formed, with the Editors, the Organising Committee of the Conference.

Dr. D.J. Wright, for help with the administration of the Conference.

Mrs. Rita Walker, Mrs. Lyn Barker and Mrs. Val Siviter, for their excellent typing and patience with many drafts of some articles.

Dr. F.M. Burrows, for his artistic skills in the illustrations.

The Science and Engineering Research Council, for support of T.L. Thickstun for 1978-81 (under grants GR/A/5561.5, GR/B/4889.5 D18).

# Preface

A conference on *Low Dimensional Topology* was held at the University College of North Wales, Bangor, on July 2-5, 1979, with 63 participants.

Four principal speakers were invited:

*W. Thurston* gave four lectures on "Hyperbolic geometry and 3-manifolds" and one lecture on "The Smith conjecture".

*A. Casson* gave one lecture on "Bordisms of automorphisms of surfaces" and two lectures on "Fake $S^3 \times \mathbb{R}$".

*R.D. Edwards* gave three lectures on "The double suspension of homology 3-spheres".

*L. Siebenmann* gave two lectures on "Fake $S^3 \times \mathbb{R}$" and one lecture on "Algebraic knots".

The articles of this volume were either presented to the conference, or arose out of it. All were refereed.

Of the lectures of the principal speakers, only those of W. Thurston appear here. The notes were written up by Peter Scott, and we asked him to write a brief survey as an introduction.

The four lectures on fake $S^3 \times \mathbb{R}$ correspond to a Séminaire Bourbaki exposé (February 1979, No. 576) by Siebenmann, based on A. Casson's work and M. Freedman's article in Annals of Math. 110 (1979) 177-201.

L. Siebenmann's lecture on algebraic knots was an introduction to material in a monograph entitled "New geometric splittings of classical knots" written jointly with F. Bonahon, which will appear shortly in this series, indeed as a second volume of these proceedings.

Peter Stefan's work on Whitehead's conjecture in 1978 has been a stimulus to subsequent research, and we felt his seminal letter of May, 1978, should be published. The article by R. Brown and J. Huebschmann grew out of an attempt to give the complete background for understanding this letter.

# Addresses of contributors


Dr. F. Bonahon,
Département de Mathematique,
Université de Paris-Sud,
Centre D'Orsay,
91405 ORSAY Cedex,
FRANCE.

Dr. W.R. Brakes,
Faculty of Mathematics,
The Open University,
Walton Hall,
MILTON KEYNES,
MK7 6AA U.K.

Professor R. Brown,
School of Mathematics and
Computer Science,
University College of North Wales,
BANGOR, LL57 2UW
Gwynedd. U.K.

Dr. L. Contreras Caballero,
Department of Pure Mathematics,
Mathematics Institute,
16 Mill Lane,
CAMBRIDGE,
CB2 1SB U.K.

and

Facultad de Ciencias Matematicas,
Universidad Complutense,
Madrid-3
SPAIN.

Dr. D. Cooper,
Mathematics Institute,
University of Warwick,
COVENTRY,
CV4 7AL U.K.

Dr. J. Huebschmann,
Mathematisches Institut,
Universitat Heidelberg,
6900 HEIDELBERG 1,
Im Neuenheimer Feld 288, B.R.D.

Professor L. Kauffman,
Department of Mathematics,
University of Illinois at Chicago
Circle, Chicago,
ILLINOIS 60680,
U.S.A.

Dr. C. Kearton,
Mathematics Department,
Durham University,
Old Shire Hall,
DURHAM,
CH1 3HP U.K.

Professor P. Orlik,
Mathematics Department,
University of Wisconsin,
Madison
WISCONSIN 53706,
U.S.A.

Dr. R. Riley,
Institute for Advanced Studies,
Princetown,
NEW JERSEY 08540,
U.S.A.

Dr. P. Scott,
Pure Mathematics Department,
The University,
P.O. Box 147,
LIVERPOOL,
L69 3BX U.K.

Professor L. Siebenmann,
Département de Mathematique,
Université de Paris-Sud,
Centre D'Orsay,
91405 ORSAY Cedex,
FRANCE.

Professor W. Thurston,
Department of Mathematics,
University of Princeton,
Princeton,
NEW JERSEY 08544,
U.S.A.

UNIVERSITY OF WALES/ PRIFYSGOL CYMRU

# UNIVERSITY COLLEGE OF NORTH WALES

# CONFERENCE ON LOW-DIMENSIONAL TOPOLOGY

## JULY 2ND – 6TH 1979

ORGANISING COMMITTEE

R. BROWN, P. SCOTT, T. L. THICKSTUN

ACCOMMODATION AVAILABLE IN A HALL OF RESIDENCE

JULY 1ST – 6TH

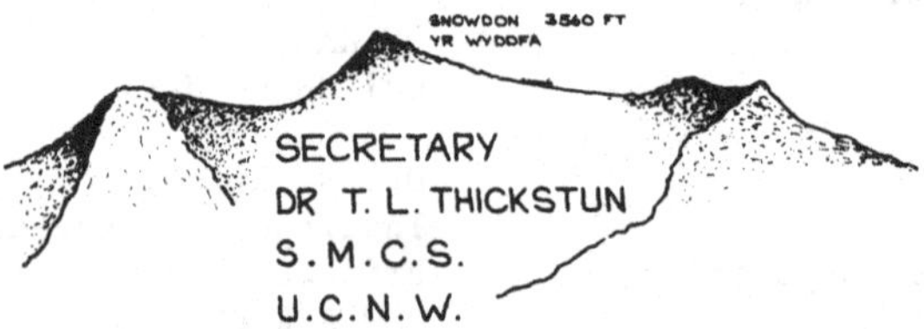


SECRETARY
DR T. L. THICKSTUN
S.M.C.S.
U.C.N.W.
BANGOR
GWYNEDD LL57 2UW
U.K.

FMB

# Part 1: 3-manifolds

# The classification of compact 3-manifolds

P. SCOTT

The aim of this article is to summarise the basic facts known about the classification of compact 3-manifolds, and to explain briefly how Thurston's recent work fits in and adds to these facts. Essentially, this note forms the background needed to appreciate the conjecture at the beginning of my notes of Thurston's lectures (which follow this article). Hempel's book [1] is the reference for all results stated here for which a specific reference is not given.

Our starting point is a theorem of Kneser proved in the 1930's, about connected sums of compact 3-manifolds. Call a 3-manifold $M$ *prime* if for any decomposition $M = M_1 \# M_2$ as a connected sum, one of $M_1, M_2$ is homeomorphic to $S^3$. Kneser proved that a compact 3-manifold can always be expressed as a connected sum of prime manifolds. Milnor added to this the information that if $M$ is also orientable, then the prime summands of $M$ are unique. If $M$ is non-orientable, this need not be the case but the only examples here are similar to those in surface theory. For example if $F$ is a non-orientable surface and $T$ and $K$ denote the torus and Klein bottle, then $F \# T$ is homeomorphic to $F \# K$. Similarly, if $F$ is a non-orientable 3-manifold, and $S^1 \tilde{\times} S^2$ and $S^1 \times S^2$ denote the non-trivial and trivial $S^2$-bundles over $S^1$, then $N \# (S^1 \tilde{\times} S^2)$ is homeomorphic to $N \# (S^1 \times S^2)$.

The preceding paragraph shows that one can reduce the classification problem for compact 3-manifolds to that for prime manifolds. At this point, one needs to make another definition. We will say that a 3-manifold $M$ is *irreducible* if any embedded 2-sphere in $M$ bounds a 3-ball. Clearly an irreducible manifold is prime. The converse is not true but the only 3-manifolds which are prime and not irreducible are the two $S^2$-bundles over $S^1$ mentioned above. Thus we can restrict our attention to irreducible manifolds.

If $M$ is orientable and irreducible, then the Sphere Theorem at once implies that $\pi_2(M) = 0$. It now follows easily that the universal covering of $M$ is contractible or is a homotopy 3-sphere $\Sigma$. From this, it follows that $M$ is aspherical with infinite, torsion free fundamental group, or $M$ is $D^3$, or $M$ is finitely covered by a homotopy 3-sphere. This last possibility gives rise to many extremely hard problems of which the Poincaré Conjecture is just one.

If $M$ is non-orientable and is $P^2$-irreducible, i.e. $M$ is irreducible and contains no two-sided projective planes, then the Projective Plane Theorem implies that $\pi_2(M) = 0$, so that again the universal covering of $M$ is contractible or is a homotopy 3-sphere $\Sigma$. The second case cannot occur as any orientation reversing homeomorphism of $\Sigma$ must have a fixed point and so cannot be a covering transformation.

Finally if $M$ is non-orientable and irreducible but does admit two-sided projective planes, then there is a canonical family of disjoint, non-parallel, two-sided projective planes in $M$ which cut $M$ into pieces in each of which any two-sided projective plane must be parallel to a boundary component. Thus all the pieces we obtain have at least one boundary projective plane. If we take the orientable double covering of such a piece and glue a 3-ball to each boundary sphere, the resulting manifold $M_1$ will be irreducible.

We conclude from the above that any compact 3-manifold can be cut up by 2-spheres and projective planes in a way which is very nearly canonical into pieces which after gluing 3-balls to all boundary spheres, cannot be further cut up. Note that many of these pieces are aspherical. This fact explains the very close connection between 3-dimensional topology and combinatorial group theory.

We now consider the compact orientable irreducible case. If $\pi_1(M)$ is finite, $M$ is the 3-ball or is covered by a homotopy 3-sphere. Apart from the 3-ball, all known examples are Seifert fibre spaces. If the Poincaré Conjecture holds and if all free finite group action on $S^3$ are equivalent to standard actions, then these examples are the only possible examples. On the other hand when $\pi_1(M)$ is infinite, Seifert fibre spaces are rather unusual. For example, the only knots whose complements are Seifert fibred are torus knots. A very nice class of manifolds to consider is the class of Haken manifolds. Recall that a surface $F$, not $S^2$, properly embedded in a 3-manifold $M$ in a two-sided manner, is *incompressible* if the induced map

$\pi_1(F) \to \pi_1(M)$ is injective. A manifold $M$ is *Haken* if there is a sequence of 3-manifolds $M = M_0, M_1, \ldots, M_n$ and surfaces $F_0, F_1, \ldots, F_{n-1}$ such that $F_i$ is an incompressible surface in $M_i$, $M_{i+1}$ is obtained from $M_i$ by cutting along $F_i$ and $M_n$ is a disjoint union of 3-balls. This sequence is called a *hierarchy*. Waldhausen proved that any compact, orientable, irreducible 3-manifold with non-empty boundary is Haken. However there are many examples of closed, orientable, irreducible 3-manifolds with infinite fundamental group which are not Haken. The known examples are all Seifert fibre spaces or have a complete hyperbolic structure so they do not provide counter examples to Thurston's conjecture. One of the specially nice features of Haken manifolds is that they tend to be characterised by their fundamental group. For closed Haken manifolds this is exactly true, i.e. closed Haken manifolds with isomorphic fundamental groups are homeomorphic. For manifolds with boundary, there are analogous results but they are more complicated to state. In a sense, this reduces the classification problem for Haken manifolds to group theory, but in general it is hard to tell if two groups are isomorphic and there is also no useful group theoretic characterisation of the fundamental groups of Haken manifolds.

As I said before, there are examples of closed orientable irreducible 3-manifolds with infinite fundamental group which are not Haken. However, the Seifert fibre space examples are known to have a finite covering space which is Haken. This led Waldhausen to the following

CONJECTURE *Any closed orientable irreducible 3-manifold with infinite fundamental group has a finite covering space which is Haken.*

This conjecture holds for some of the known hyperbolic non-Haken examples, but has not been proved for all the known examples. It is not even known if all of these examples have closed surface groups contained in their fundamental group. In the general case, when a hyperbolic structure is not assumed, it is not even known that the universal covering space $\tilde{M}$ of a non-Haken manifold $M$ must be homeomorphic to $\mathbb{R}^3$. The main problem here is that the ends of the universal covering space are not known to be tame. Until recently, there was also the problem that $\tilde{M}$ might contain fake balls. But this has now been shown to be impossible by Yau using minimal surface theory.

The class of Haken manifolds is extremely large but the sub-class of Haken Seifert fibre spaces is small and completely under-stood. The work of Johannson [3] and of Jaco and Shalen [2]

shows that any Haken manifold has a Seifert fibred part and a non-Seifert fibred part. In fact they proved much more, but for the purposes of this note we shall concentrate on this particular result. If a compact Haken manifold M has compressible boundary, then the Loop Theorem gives a 2-disk D properly embedded in M. Let N be the manifold obtained by cutting M along D. Thus M is obtained from N by attaching a 1-handle to $\partial N$. This simple description allows us to consider only Haken manifolds with incompressible boundary. All other Haken manifolds are obtained by adding 1-handles. Johannson [3] and Jaco and Shalen [2] show that if M is a Haken manifold with incompressible boundary, then there is a canonical family of disjoint incompressible tori which cuts M into pieces which are either Seifert fibred spaces or are atoroidal. The atoroidal Haken manifolds are exactly the manifolds whose interiors Thurston tells us admit a complete hyperbolic structure.

The preceding discussion of Haken manifolds was for the orientable case only. In the non-orientable case the situation is easier in some respects. In particular any non-orientable, compact, $P^2$-irreducible 3-manifold is Haken. As in the orientable case, one can also restrict attention to manifolds with incompressible boundary. The orientable double covering $\tilde{M}$ of such a manifold M is automatically Haken and will again have incompressible boundary. Hence the Torus Splitting Theorem mentioned above holds for $\tilde{M}$. Now it is not hard to deduce that the canonical family of tori in $\tilde{M}$ can be isotoped to the invariant under the covering involution. Hence, M has a canonical family of tori and Klein bottles which cuts M into pieces which are either atoroidal or are double covered by Seifert fibre spaces. The conclusion of the preceding paragraphs taken together with Thurston's hyperbolisation theorem is that any Haken manifold can be cut canonically into Seifert fibre space pieces and hyperbolic pieces. Now the fundamental group of a hyperbolic piece is a sub-group of $PSL(2, \mathbb{C})$ and this gives much new information about 3-manifold groups. For example, it follows that the fundamental group of a Haken manifold is residually finite. But even more importantly, the hyperbolisation result tells one to use geometric ideas as opposed to purely topological ideas when considering 3-manifolds. For example, the question of when a covering space of a Haken manifold can be compactified has now been answered by Thurston in many cases. Thus, although it is not clear that a classification of 3-manifolds is closer, Thurston's work has already had a great impact on 3-dimensional topology and on Kleinian group theory, and is going to lead to a much deeper understanding of both subjects.

## REFERENCES

1. J. HEMPEL, '3-manifolds', *Ann. of Math. Studies 86*, Princeton University Press (1976).

2. W. JACO and P.B. SHALEN, 'Seifert fibred spaces in 3-manifolds' *Memoir of American Math. Soc.* No. 220.

3. K. JOHANNSON, 'Homotopy equivalences of 3-manifolds with boundaries', *Springer Lecture Notes* 761, 1979.

# Hyperbolic geometry and 3-manifolds

W. THURSTON

My theme is that geometrical methods yield more information about manifolds than do purely topological methods and in dimension three geometric methods are often applicable. A good example of this is the recent proof of the Smith Conjecture.

Here is the strongest possible conjecture asserting that one can always use geometry when studying 3-manifolds. The Poincaré Conjecture is a very special case.

CONJECTURE. *Every compact 3-manifold* $M$ *with incompressible boundary has a canonical decomposition into geometric pieces, i.e. by cutting* $M$ *along a canonical family of disjoint, 2-sided, closed surfaces each homeomorphic to* $S^2$, $P^2$, $T^2$ *or the Klein bottle, one can obtain geometric pieces.*

When I say that a 3-manifold $M$ is *geometric*, I mean that the interior of $M$ has a complete geometric structure modelled on some homogeneous space. By a *homogeneous space*, I shall mean a space $X$ and a transitive group $G$ of homeomorphisms of $X$ with the property that $G_x$, the stabiliser of $x$, is compact for every $x$ in $X$. It follows that $X$ admits a G-invariant metric. We will always assume that $X$ is equipped with such a metric and we will usually assume that $G$ is maximal i.e. the full isometry group of $X$.

A manifold $M$ without boundary has a $(X,G)$-structure if it is locally homeomorphic to open subsets of $X$ and there is an atlas of charts such that all the overlap maps lie in $G$. Such a manifold inherits a metric from the metric on $X$.

If $X$ is simply connected and $M$ has a complete $(X,G)$-structure, then $M$ is isometric with the quotient of $X$ by some subgroup $\Gamma$ of $G$ acting as a group of covering transformations of $X$. In particular $\Gamma$ is isomorphic to $\pi_1(M)$. Of course, a homogeneous space $(X,G)$, and hence any manifold with $(X,G)$-

structure, has constant Gauss curvature.

The following result is a partial affirmation of the conjecture.

THEOREM 1. *The conjecture holds for Haken manifolds.*

The more usual formulation of the result is that one can cut an orientable Haken manifold with incompressible boundary along tori to obtain pieces which are Seifert fibre spaces or admit a complete hyperbolic structure. This gives a somewhat misleading picture as there are many relevant geometries apart from hyperbolic geometry. It just happens that most of those geometries give rise to Seifert fibre spaces.

Note that if $M$ is a Haken manifold with compressible boundary one can cut $M$ along 2-discs to obtain a Haken manifold with incompressible boundary. Thus $M$ also admits a decomposition into geometric pieces. However, in general, this decomposition is not canonical as the family of 2-discs is not canonical, e.g. if $M$ is a handlebody.

The first step in approaching the conjecture is to understand which types of geometry can occur. Thus one wants to classify simply connected Riemannian manifolds with transitive isometry groups, and hence constant Gauss curvature.

In dimension 2, it is easy to classify simply connected homogeneous spaces by their Gauss curvature $K$. By an appropriate scale change one can assume that $K$ is $-1, 0$ or $1$. The three corresponding homogeneous spaces are $H^2, E^2$ and $S^2$. A nice fact here is that the isometry group is transitive on directions as well as points, so that the three spaces have constant curvature, i.e. constant sectional curvature. This does not hold in dimension three. In the case $K = 0$, the torus is the only closed orientable surface admitting a $(E^2, \mathrm{Isom}(E^2))$ structure. These structures give the flat tori. We now fix a torus $T$ and an element $\alpha$ of $\pi_1(T)$ which can be represented by an embedded loop in $T$. Given a flat structure on $T$, we first make a scale change so that the area of $T$ is $1$. There is a closed geodesic $\lambda$ (which is not unique) which represents $\alpha$. Cut $T$ along $\lambda$ and let $\ell$ be the length of $\lambda$. One obtains a cylinder $C$ of height $\ell^{-1}$ and circumference $\ell$. This cylinder is completely determined up to isometry by the number $\ell$. Hence the flat structure on $T$ is determined by $\ell$ and by a real number $\theta$ which measures the twist used to glue the two boundary circles of $C$. Hence the space of flat structures on $T$ is homeomorphic to $\mathbb{R}^2$, with a particular structure having coordinates $(\log \ell, \theta)$.

In the case $K < 0$, there is a similar description for the space of closed orientable surfaces of a given genus with hyperbolic structure. This is the Teichmüller space of the surface. One cuts the surface of genus $g$, $M$ say, along $3g - 3$ closed geodesics in the homotopy classes shown in Fig. 1 so that $M$ falls into $2g - 2$ pairs of pants. Each pair of pants is determined by the lengths of its three boundary curves, and $M$ is determined by these lengths together with a twist parameter for each of the circles. Thus the Teichmüller space of $M$ has dimension $6g - 6$.

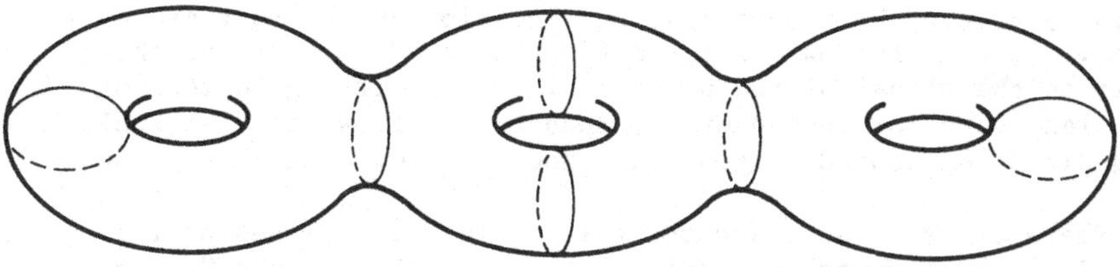

Fig. 1

Now we turn to the 3-dimensional homogeneous spaces. There are eight relevant simply connected ones. *Relevant* means that there is a manifold with a complete (X,G)-structure which has finite volume. An example of an irrelevant homogeneous space is obtained by taking $X$ to be $\mathbb{R}^3$ and $G$ to be generated by homeomorphisms of the form

$$(x,y,z) \to (x+r,\ y+s,\ z)\ ,\quad \forall r,s \in \mathbb{R}$$

$$(x,y,z) \to (e^{-t}x,\ e^{-t}y,\ z+t)\ ,\ \forall t \in \mathbb{R}\ .$$

We obtain the classification of the 3-dimensional homogeneous spaces by considering the size of $G_x$. For convenience we will take $G$ to be a group of orientation preserving isometries.

<u>Case $G_x = SO_3$</u>

In this case $X$ has constant curvature and, as in the 2-dimensional case, we obtain the examples $H^3$, $E^3$ and $S^3$ with curvature $-1, 0, 1$. In each case, if $G$ denotes the full orientation preserving isometry group of $X$, there is a bundle $SO_3 \to G \to X$.

Case $G_x = SO_2$

Let $T_x$ be the tangent space at $x$. Then $T_x = L_x \oplus P_x$ where $L_x$ is the line fixed by $G_x$ and $P_x$ is the orthogonal plane which is invariant under $G_x$. As $X$ is simply connected, we can choose coherent orientations on the $L_x$, so as to obtain a unit vector field on $X$. This vector field and the plane field $P_x$ are both G-invariant and so descend to any manifold covered by $X$. The flow on $X$ determined by the vector field also leaves the plane field invariant. This flow must preserve the area of the planes. This is because $X$ has a quotient $M$ of finite volume, by hypothesis, and the flow on $M$ must preserve volume. It then follows that the flow on $X$ must preserve the metric on the planes of the plane field. For if some direction in a plane were contracted or expanded by the flow, then so would every direction expand or contract, and the flow could not preserve area.

There are six cases according to whether the planes of the plane field have Gauss curvature $k > 0$, $k = 0$ or $k < 0$ and according to whether the plane field is integrable or not. Two of the cases, $S^3$ and $E^3$, have appeared already, so we have a total of seven homogeneous spaces so far. The six cases are

| | $k > 0$ | $k = 0$ | $k < 0$ |
|---|---|---|---|
| Integrable | $S^2 \times E$ | $E^2 \times E^1 = E^3$ | $H^2 \times E$ |
| Not integrable | $S^3$ | $N$ | $\widetilde{SL_2(\mathbb{R})}$ |

where $N$ is the nilpotent Lie group of dimension three and $\widetilde{SL_2(\mathbb{R})}$ is the universal cover of $SL_2(\mathbb{R})$. There is an exact sequence $0 \to \mathbb{R}^1 \to N \to \mathbb{R}^2 \to 0$, so the four spaces in the case $k \le 0$ are all topologically $\mathbb{R}^3$. The space $N$ gives the correct geometry for non-trivial circle bundles over the torus.

Case $G_x = 1$

In this case, $X = G$ and we have a Lie group. The only new possibility is the solvable Lie group $S$ given as a split extension $0 \to \mathbb{R}^2 \to S \to R \to 0$, where $t$ in $\mathbb{R}$ acts on $\mathbb{R}^2$ by the matrix $\begin{pmatrix} e^t & 0 \\ 0 & e^{-t} \end{pmatrix}$.

The compact 3-manifolds obtained from this homogeneous space are all torus bundles over $S^1$. (See Ch.4 of [4].)

Observe that all the compact, orientable manifolds whose interiors admit complete geometric structures modelled on $S^3$, $E^3$, $S^2 \times E$, $H^2 \times E$, $\widetilde{SL_2(\mathbb{R})}$ and N are Seifert fibre spaces. Thus a geometric manifold is a Seifert fibre space or a torus bundle over $S^1$ or admits a hyperbolic structure.

Now I want to talk about the result that 'most' Haken manifolds admit a complete hyperbolic structure. Theorem 1 then follows by applying the Splitting Theorem of Johannson and of Jaco and Shalen and by proving some comparatively simple facts about Seifert fibre spaces.

*Definition.* A compact 3-manifold M is *atoroidal* if every subgroup of $\pi_1(M)$ isomorphic to $\mathbb{Z} \times \mathbb{Z}$ is conjugate into the image under the inclusion map of the fundamental group of a boundary torus of M.

If M has a complete hyperbolic structure then M must be atoroidal.

THEOREM 2. *Let* M *be a compact Haken manifold which contains some closed incompressible surface which is not the fibre of a fibreing of* M *over* $S^1$. *Then* $\mathring{M}$ *has a complete hyperbolic structure if and only if* M *is atoroidal.*

REMARK. $\mathring{M}$ will have finite volume if and only if $\partial M$ is a union of incompressible tori and M is not an I-bundle over $T^2$ or the Klein bottle.

In dimension $\geq 3$, one has the

Mostow Rigidity Theorem

If $M_1^n$, $M_2^n$ each have complete hyperbolic structures of finite volume and if $f : M_1 \to M_2$ is a homotopy equivalence, then f is homotopic to an isometry.

In particular $M_1$ admits essentially only one complete hyperbolic structure.

Now suppose that M is a compact 3-manifold whose boundary is not a union of tori and that $\mathring{M}$ admits a complete hyperbolic structure. Let F be a component of $\partial M$ which is not a torus.

It is sometimes, but not always, possible to find in $\mathring{M}$ a totally geodesic surface parallel to F. Suppose that we can do this for all non-torus components of M. Let DM denote the manifold obtained by doubling M along the non-torus components of $\partial M$. Then Int (DM) is obtained by cutting $\mathring{M}$ along all the totally geodesic surfaces obtained above and doubling the resulting truncated manifold. Thus Int (DM) admits a complete hyperbolic structure and so is atoroidal. Hence M iself must be annulus-free i.e. any singular incompressible annulus A in M is properly homotopic into $\partial M$.

In fact, one can use Theorem 2 to prove the following result.

THEOREM 3. *Let* M *be a compact 3-manifold such that* $\mathring{M}$ *admits a complete hyperbolic structure. There is a totally geodesic surface in* $\mathring{M}$ *for each non-torus component of* $\partial M$ *if and only if* M *is annulus-free.*

*Proof.* We have already shown that if there is a totally geodesic surface in $\mathring{M}$ for each non-torus component of $\partial M$, then M must be annulus-free. Now suppose that M is annulus-free, so that we need to demonstrate the existence of some totally geodesic surfaces. Again let DM denote the manifold obtained by doubling M along the non-torus components of $\partial M$. As $\mathring{M}$ admits a complete hyperbolic structure, M must be atoroidal. As M is also annulus-free, it follows that DM must be atoroidal. Theorem 2 then states that DM admits a complete hyperbolic structure. This structure must have finite volume as $\partial DM$ contains only tori. There is an obvious involution $\tau$ on DM which interchanges the two copies of M and fixes their intersection. The Mostow Rigidity Theorem assures us that $\tau$ is homotopic to an isometry $\tau'$ which must still be an involution. Now it follows that for each non-torus component F of $\partial M$, there is a surface F' parallel to F and left fixed by $\tau'$. Such a surface must be totally geodesic, and this proves the required result.

Note that if a compact 3-manifold M has boundary which is not a union of tori, then $\mathring{M}$ must have infinite volume for any complete hyperbolic structure. However, there is still a rigidity theorom if M is annulus free, proved by applying the Mostow Rigidity Theorem to Int (DM). Let $M_1$, $M_2$ be two copies of M each with a complete hyperbolic structure on its interior. A homeomorphism $M_1 \to M_2$ extends to give a homeomorphism $DM_1 \to DM_2$. Mostow's Theorem implies that the homeomorphism Int $(DM_1) \to$ Int $(DM_2)$ is homotopic to an isometry, for Int $(DM_1)$ and Int $(DM_2)$ have finite volume as $\partial DM_1$ and $\partial DM_2$ are unions of tori. Now Int $(DM_1)$ and Int $(DM_2)$ each have an involution which swaps the two copies of M and is an isometry. Using our isometry between Int $(DM_1)$ and Int $(DM_2)$, we obtain two involutions of Int $(DM_1)$ which are isometries and are homotopic to each other. But homotopic isometries of Int $(DM_1)$ must be equal. It follows that there is an isometry $M_1 \to M_2$.

Note that it is a general fact that homotopic isometries of a hyperbolic 3-manifold with non-abelian fundamental group must be equal. The fact that Int $(DM_1)$ has finite volume is irrelevant. The reason is that if $\Gamma$ is a discrete subgroup of $PSL(2, \mathbb{C})$ and one has an isometry $\alpha$ of $H^3/\Gamma$, then there is a covering isometry g of $H^3$. If $\alpha$ is homotopic to the identity, then some choice of g must centralise $\Gamma$. Now the centraliser of any non-trivial element of $PSL(2, \mathbb{C})$ is abelian. Hence g must be trivial and $\alpha$ must equal the identity, so long as $\Gamma$ is not abelian.

I shall use the term Kleinian group for any discrete subgroup of $PSL(2, \mathbb{C})$, the orientation preserving isometry group of $H^3$. We shall think of $H^3$ in the Poincaré ball model, so that the action of $PSL(2, \mathbb{C})$ on $H^3$ extends to an action on $S^2$.

If $\Gamma$ is a Kleinian group, we define $L_\Gamma$, the limit set of $\Gamma$, to be the set of all limit points in $H^3 \cup S^2$ of the orbit under $\Gamma$ of a point x of $H^3$. As $\Gamma$ is discrete, $L_\Gamma$ must lie in $S^2$. Also $L_\Gamma$ is independent of the choice of the point x. If $\Gamma$ fixes no point of $S^2$, then $L_\Gamma$ is also the set of limit points of the orbit under $\Gamma$ of any point of $S^2$. One of the crucial facts about $L_\Gamma$ is that $\Gamma$ acts "very transitively" on $L_\Gamma$.

Let $D_\Gamma = S^2 - L_\Gamma$. Then $\Gamma$ acts properly discontinuously on $D_\Gamma$ and on $H^3 \cup D_\Gamma$. If $\Gamma$ is torsion free, the projection map $H^3 \cup D_\Gamma \to (H^3 \cup D_\Gamma)/\Gamma$ is a covering and the quotient is a manifold M (with boundary, if $D_\Gamma \neq \emptyset$) whose fundamental group is $\Gamma$ and whose interior has a complete hyperbolic structure. The boundary of M, which is $D_\Gamma/\Gamma$, has a natural conformal structure as $\Gamma$ acts conformally on $S^2$ and $D_\Gamma$ is an open subset of $S^2$. A point to notice here is that it is quite possible for $M$ to be non-compact, even if $\mathring{M}$ is the interior of a compact manifold $W$. The general picture, in topological terms is that $\partial M$ is obtained from $\partial W$ by first removing a finite collection of disjoint, embedded essential circles and then removing some (possible none or possibly all) of the components of the remainder. The Ahlfors Finiteness Theorem states the more general result that for any finitely generated Kleinian group $\Gamma$, the quotient surface $D_\Gamma/\Gamma$ is conformally equivalent to a finite number of finitely punctured closed surfaces.

In the general situation $\mathring{M}$ need not be the interior of a compact manifold and $\Gamma$ need not even be finitely generated, though it usually will be. However, one can always define the *convex core* of $M$. This can be described as the minimal convex submanifold of $M$, where a set $X$ in $M$ is convex if given $x_0$ and $x_1$ in $X$ and any geodesic segment $\ell$ from $x_0$ to $x_1$, then $\ell$ lies in $X$. Thus if $X$ is a non-empty convex subset of $M$, the natural map $\pi_1(X) \longrightarrow \pi_1(M)$ is surjective. One can also consider in $H^3$ the intersection of all half spaces in $H^3$ whose closure in $H^3 \cup S^2$ contains $L_\Gamma$. This is a $\Gamma$-invariant convex set in $H^3$ and its quotient by $\Gamma$ is the convex core of $\mathring{M}$.

The convex core of $\mathring{M}$ is empty when $\pi_1(M)$ consists entirely of parabolic isometries of $H^3$. Such a group must be isomorphic to $\mathbb{Z}$ or $\mathbb{Z} \times \mathbb{Z}$. Excluding this special case, the convex core of $\mathring{M}$ is always a deformation retract of $\mathring{M}$, but it need not be compact in general. Note that if $\mathring{M}$ has finite volume, then $\mathring{M}$ equals its convex core.

The proof of Theorem 2 is by induction on the length of a hierarchy for $M$. By a kind of doubling procedure (really one considers orbifolds), we can arrange that we only need to consider closed surfaces in our hierarchy.

Now we come to one of the key results for the proof of Theorem 2. This is the 3-dimensional analogue of the fact that there is a bijection between hyperbolic structures on a pair of pants and the lengths of the three boundary curves of the pair of pants.

*Definition* Let $X$ and $Y$ be metric spaces. Then $f : X \to Y$

is a *quasi-isometry* if there exists $K, L$ with

$$0 < K \leq \frac{d(x_1,x_2)}{d(fx_1,fx_2)} \leq L ,$$

for all $x_1 \neq x_2$ in $X$.

*Definition* If $\Gamma$ is a Kleinian group, then a *quasi-isometric deformation* of $\Gamma$ is a group $\Gamma'$ quasi-isometrically conjugate to $\Gamma$ in its action on $H^3$ i.e. $H^3/\Gamma$ is quasi-isometric with $H^3/\Gamma'$. (In particular, $H^3/\Gamma$ is homeomorphic to $H^3/\Gamma'$, and $D_\Gamma$ is homeomorphic to $D_{\Gamma'}$).

THEOREM 4 (Ahlfors, Bers, ..., Sullivan)

*Let* $\Gamma$ *be a finitely generated Kleinian group. Then the quasi-isometric deformations of* $\Gamma$ *correspond bijectively with conformal structures on* $D_\Gamma/\Gamma$.

There is a natural map from quasi-isometric deformations of $\Gamma$ to conformal structures on $D_\Gamma/\Gamma$. This map was proved to be a surjection by Ahlfors and Bers. The idea is as follows. A conformal structure on $D_\Gamma/\Gamma$ induces one on $D_\Gamma$ itself. Now the Riemann mapping theorem implies that there is a conformal equivalence of $S^2$ sending $D_\Gamma$ with a given conformal structure to $D_\Gamma$ with the conformal structure coming from $\Gamma$. This equivalence of $S^2$ can be extended to a quasi-isometry of $H^3$ in such a way as to obtain a quasi-isometric deformation $\Gamma'$ of $\Gamma$ inducing the given conformal structure on $D_\Gamma/\Gamma$.

Sullivan showed that the natural map from quasi-isometric deformations of $\Gamma$ to conformal structures on $D_\Gamma/\Gamma$ is also injective.

Now we consider what is the basic problem in the induction step of the proof of Theorem 2. One has two compact manifolds $M_1$ and $M_2$ each with a complete hyperbolic structure on its interior. One has incompressible components $F_1$ and $F_2$, not tori, of $\partial M_1$ and $\partial M_2$ respectively and wishes to show that the manifold $M$ obtained by gluing $M_1$ to $M_2$ by a homeomorphism $\phi : F_1 \to F_2$ also admits a complete hyperbolic structure. Clearly Theorem 4 has some relevance to this problem, but a more careful look at the problem shows that we need much more information. Basically, the conformal structure on $F_1$ is "at infinity" in terms of the hyperbolic structure on $\mathring{M}_1$ and it does not make sense metrically to glue $\mathring{M}_1 \cup F_1$ to $\mathring{M}_2 \cup F_2$. We would like to find in $\mathring{M}_1$ a surface $L_1$ parallel to $F_1$. If we could find a totally geodesic surface $L_1$ parallel to $F_1$ and if we could find a totally geodesic surface $L_2$ in $M_2$ parallel to $F_2$ we could arrange that $L_1$ and $L_2$ are isometric by Theorem 4 and then we could simply cut $\mathring{M}_1$ at $L_1$, cut $\mathring{M}_2$ at $L_2$ and glue $L_1$ to $L_2$ to obtain a hyperbolic manifold homeomorphic to $M$. However, if $M_1$ is not annulus free this may be impossible. The best one can do is to find a surface $L_1$ parallel to $F_1$ such that when we cut off the region between $L_1$ and $F_1$, we obtain a convex submanifold of $M_1$ as in Fig. 2.

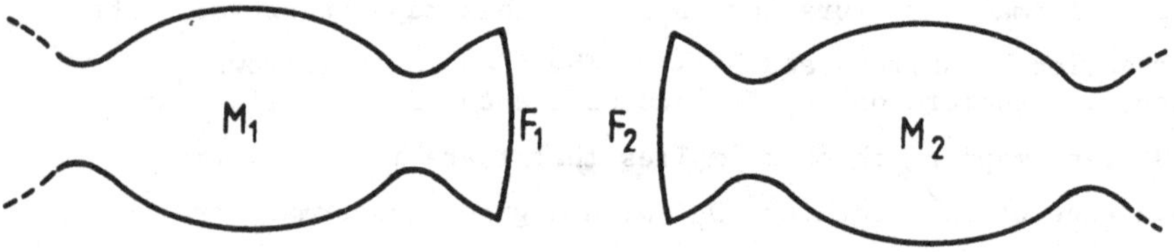


Fig. 2

Of course, in Fig. 2, one cannot glue $L_1$ to $L_2$ as they are "convex in opposite directions".

We need to consider quasi-Fuchsian groups. These are abstractly isomorphic to Fuchsian groups but their limit set is a Jordan curve in $S^2$, whereas a Fuchsian group has limit set a round circle in $S^2$. The quotient of $H^3$ by a torsion free, quasi-Fuchsian group is topologically $F \times \mathbb{R}$ where $F$ is a surface and one has two associated conformal structures on $F$, one at each end of $F \times \mathbb{R}$.

Let $M$ be a compact 3-manifold whose interior admits a complete hyperbolic structure such that $D_\Gamma/\Gamma$ equals $\partial M$. For convenience, suppose that $\partial M$ is connected, incompressible and is not a torus. Then we will define a map

$$\sigma_M : T(\partial M) \longrightarrow T(\partial M)$$

where $T(\partial M)$ is the Teichmuller space of $\partial M$.

Let $\tilde{M}$ be the covering of $M$ with $\pi_1(\tilde{M}) = \pi_1(\partial M)$. The group $\pi_1(\partial M)$ is quasi-Fuchsian because $\partial M = D_\Gamma/\Gamma$, so the quasi-isometric deformations of $\pi_1(\tilde{M})$ correspond to the conformal structures on $\partial M$ at the two ends of $\mathring{\tilde{M}}$. As $\partial M$ lifts to $\tilde{M}$, one of the structures will be the same as the original structure on $\partial M$. Hence we have the following diagram which defines $\sigma_M$.

$$\begin{array}{ccccc} T(\partial M) & & \xrightarrow{\quad\quad\sigma_M\quad\quad} & & T(\partial M) \\ \downarrow \cong & & & & \uparrow P_2 \\ \text{quasi-isometric deformation of } M & \longrightarrow & \text{quasi-isometric deformations of } \tilde{M} & \xrightarrow{\cong} & T(\partial M) \times T(\partial M) \end{array}$$

Here $P_2$ denotes projection of $T(\partial M) \times T(\partial M)$ onto the second factor, where the factors are ordered so that the first factor of $T(\partial M) \times T(\partial M)$ simply contains the original conformal structure on $\partial M$.

Now we return to $(M_1, F_1)$ and $(M_2, F_2)$. The surface $F_1$ has two conformal structures, the obvious one which we denote by $g$ and the other one $\sigma_{M_1}(g)$. Similarly we have conformal structures $h$, $\sigma_{M_2}(h)$ on $F_2$. We are given $\phi : F_1 \to F_2$ and require that

$$\phi_*(\sigma_{M_1}(g)) = h, \phi_*(g) = \sigma_{M_2}(h) ,$$

i.e. we want $\phi_*(g) = \sigma_{M_2}(h) = \sigma_{M_2}\phi_*\sigma_{M_1}(g)$, or, equivalently, $g = \phi_*^{-1}\sigma_{M_2}\phi_*\sigma_{M_1}(g)$.

So we need to know that the map $\phi_*^{-1}\sigma_{M_2}\phi_*\sigma_{M_1} : T(\partial M_1) \to T(\partial M_1)$ has a fixed point. If this happens, it is then possible to glue

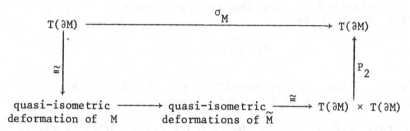

$M_1$ to $M_2$ using the Maskit Combination Theorems. We have representations of $\pi_1(M_1)$ and $\pi_1(M_2)$ in $PSL(2, \mathbb{C})$ and the existence of the fixed point implies that we can conjugate one representation so that they agree on $\pi_1(F_1)$. This is because quasi-Fuchsian representations of surface groups are determined up to conjugacy by the conformal structures at the two ends of the quotient manifolds, by Theorem 4. We now have a representation of

$$\pi_1(M_1) *_{\pi_1(F_1)} \pi_1(M_2)$$

in $PSL(2, \mathbb{C})$ and the Combination Theorems imply that this representation is discrete and faithful. Waldhausen's results on Haken manifolds imply that the quotient hyperbolic manifold must be homeomorphic to

$$M_1 \cup_{F_1} M_2 ,$$

because it has the same fundamental group. So we have obtained a hyperbolic structure on this union, as required.

We know that if $M_1$ and $M_2$ are not annulus free, it may be impossible to glue $M_1$ to $M_2$ and obtain a complete hyperbolic structure on the result. For example, if $M_1 = M_2$ and the gluing map gives the double of $M_1$, then the new manifold is not atoroidal. Thus the existence of a fixed point for the map $\phi_*^{-1}\sigma_{M_2}\phi_*\sigma_{M_1} : T(\partial M_1) \to T(\partial M_1)$ must, in general, depend on the homeomorphism $\phi$. However, in the annulus-free case we have

THEOREM 5 *If* $M$ *is annulus-free,* $\mathring{M}$ *admits a complete hyperbolic structure and* $\partial M$ *is connected and not a torus, then the image of* $\sigma_M : T(\partial M) \to T(\partial M)$ *is bounded i.e. has compact closure.*

As $T(\partial M)$ is homeomorphic to Euclidean space of some dimension, the existence of a fixed point for the map $\phi_*^{-1}\sigma_{M_2}\phi_*\sigma_{M_1}$ follows by the Brouwer fixed point theorem.

The idea of the proof of Theorem 5 is that there is a natural compactification of $T(\partial M)$ and that $\sigma_M$ extends to the compactification. We will return to this point later.

I want to discuss in a bit more detail the picture 'at infinity' in the universal covering of a hyperbolic 3-manifold. Let $M$ be a compact Haken manifold which is annulus-free and has incompressible boundary (possibly disconnected). For convenience we will also assume that $M$ has no boundary tori. Suppose that $M$ has a complete hyperbolic structure so that $\mathring{M} = H^3/\Gamma$ , where $\Gamma$ is a discrete subgroup of $PSL(2, \mathbb{C})$ . Suppose also that $D_\Gamma/\Gamma$ equals $\partial M$ , so that $(H^3 \cup D_\Gamma)/\Gamma$ equals $M$ .

The fact that $\partial M$ is incompressible is equivalent to the statement that each component of $D_\Gamma$ is simply connected. The fact that $M$ is annulus-free is equivalent to the statement that distinct components of $D_\Gamma$ have stabilisers which intersect trivially.

Recall that there is a conformal structure associated to each component $F$ of $\partial M$ which is not a torus. The inverse image of $F$ in $H^3 \cup S^2$ , is a collection of disjoint open cells which are components of $D_\Gamma$ . The conformal structure on one of these cells say $E$ , determines that on $F$ . The complement of $\bar{E}$ in $S^2$ is another open cell. The conformal structure on this cell determines the conformal structure $\sigma_M(g)$ on $F$ . Now $S^2 - \bar{E}$ contains all components of $D_\Gamma$ apart from $E$ itself, and the component of $D_\Gamma$ in the inverse image of $F$ project to patches in $F$ each with its own conformal structure giving a leopard picture on $F$ as well as on $S^2$ . These patches on $F$ determine the new conformal structure on $F$ .

Now we return to the proof of Theorem 5. Let $M$ be a compact Haken manifold which is annulus-free and has incompressible boundary and suppose that $\mathring{M}$ admits a complete hyperbolic structure so that $\mathring{M} = H^3/\Gamma$ and $M = (H^3 \cup D\Gamma)/\Gamma$ . Let $F$ denote the set of all faithful representations $\Gamma \to PSL(2, \mathbb{C})$ which have discrete image, under the equivalence relation of conjugacy in $PSL(2, \mathbb{C})$ . If $\rho : \Gamma \to PSL(2, \mathbb{C})$ has discrete image then $H^3/\rho(\Gamma)$ is homeomorphic to $M$ , but $D_{\rho(\Gamma)}/\rho(\Gamma)$ need not equal $\partial M$ . A useful fact is that any such representation $\Gamma \to PSL(2, \mathbb{C})$ lifts to $SL(2, \mathbb{C})$ , as $M$ (in fact, any orientable 3-manifold) admits a spin structure.

There are several natural topologies on $F$ . We will use the algebraic one which comes from the compact-open topology on the set of all representations $\Gamma \to PSL(2, \mathbb{C})$ . The interior of $F$ (in the space of conjugacy classes of all representations) corresponds to the quasi-isometric deformations of $\Gamma$ and so is homeomorphic to $T(\partial M)$ , the Teichmuller space of the non-torus components of

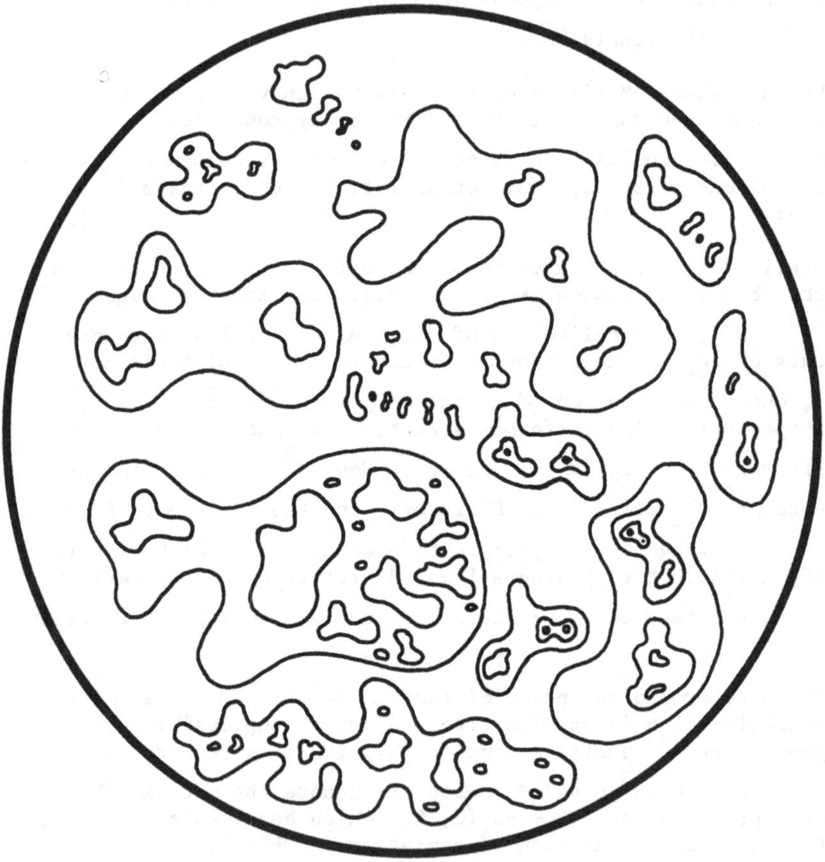

Fig. 3.

$S^2$ is cut by $L_\Gamma$ into patches which are the components of $D_\Gamma$ giving a leopard spotted picture. Usually $D_\Gamma$ has infinitely many components.

$\partial M$ , by Theorem 4. A typical example of a non-interior point of F is obtained as follows. Pick some essential embedded circle in a non-torus component of $\partial M$ , squeeze it to a point and remove the bad point so produced leaving two punctures in $\partial M$ . This conformal structure on the punctured $\partial M$ can be thought of as a limit of points of $T(\partial M)$ . (In fact, $T(\partial M)$ has a natural compactification with such structures among the extra points of the compactification.) This sequence of points of $T(\partial M)$ determines a sequence of hyperbolic structures on $\mathring{M}$ , by Theorem 4, and the limiting structure, $\rho : \Gamma \to PSL(2, \mathbb{C})$ , will have $D_{\rho(\Gamma)}/\rho(\Gamma)$ equal to $\partial M$ with two punctures.

LEMMA 6. *Let* M *be a compact, Haken manifold which is annulus-free and has incompressible boundary and suppose that* $\mathring{M}$ *admits a complete hyperbolic structure. Then the space* F *, defined above, is Hausdorff.*

THEOREM 7. *With the same hypotheses as in Lemma 6,* F *is compact.*

Note that if $\mathring{M}$ has finite volume, it follows easily from the Mostow rigidity theorem that F is finite. Theorem 7 is the generalisation of this result to the case of infinite volume.

Now recall that the interior of F is homeomorphic to $T(\partial M)$ . Identifying these two spaces one can prove

PROPOSITION 8. *The map* $\sigma_M : T(\partial M) \to T(\partial M)$ *extends to a map* $F \to T(\partial M)$ .

This proves that $\sigma_M$ has bounded image, as stated in Theorem 5.

I would like to end by giving an idea of the proof of Theorem 7, which asserts that F is compact. This is equivalent to asserting that, for fixed M , the set of equivalence classes of homotopy equivalences $M \to M'$ is compact. I will consider the special case when M is closed. Theorem 7 reduces to

THEOREM 9. *If* M *is a closed hyperbolic* 3*-manifold, then the set of homotopy classes of self homotopy equivalences is finite. Equivalently,* $Out(\pi_1(M))$ *is finite.*

As I said before, this follows easily from the Mostow rigidity theorem. Let $\Gamma$ denote $\pi_1(M)$ . An isometry of M lifts to an isometry of $H^3$ which normalises $\Gamma$ . Let $N(\Gamma)$ denote the normaliser of $\Gamma$ . The first step is to observe that as $\Gamma$ is discrete, $N(\Gamma)$ must also be discrete. This is because an element of $N(\Gamma)$ , very close to the identity, must centralise $\Gamma$ and must therefore be the identity. As $N(\Gamma)$ is discrete, $H^3/N(\Gamma)$ has a non-zero volume. As $H^3/\Gamma$ has finite volume, it follows that $\Gamma$

must have finite index in $N(\Gamma)$ . Now $Out(\Gamma)$ is exactly equal to the quotient $N(\Gamma)/\Gamma$ , by the rigidity theorem, so Theorem 9 follows.

Now I will give a more direct and geometric proof of Theorem 9 which can be generalised to give a proof of Theorem 7.

*Proof of Theorem 9:* Choose a triangulation of $M$ . This will remain unchanged throughout. If $f : M \to M$ is any map, there is a homotopic map $f' : M \to M$ which agrees with $f$ on the vertices of $M$ , sends edges to geodesic segments in $M$ and sends 2-simplexes to totally geodesic (possibly singular) triangles in $M$ . We call such a map *piecewise straight*.

If there was a bound on the length of the image of the 1-skeleton of $M$ which held for all piecewise straight maps, the result would follow at once. This is because there would be a bound on the lengths of the images of the generators of $\pi_1(M)$ . We cannot expect that there is such a bound, but a uniform bound certainly exists for the area of the image of the 2-skeleton of $M$ under a piecewise straight map. For we have a fixed number of triangles in our triangulation of $M$ , and the image of each triangle has area less that $\pi$ . Now a kind of duality allows us to deduce the result we want.

To avoid confusion between source and target, call the target manifold $N$ . We have a fixed triangulation of the source manifold $M$ and a piecewise straight map $f : M \to N$ . Choose a base point $*$ in $N$ and a finite set of embedded circles $C_i$ in $N$ through $*$ which together generate $\pi_1(N, *)$ . For each $C_i$ , choose a regular neighbourhood $U_i$ , homeomorphic to $S^1 \times D^2$ . We consider $X_i = f(\text{2-skeleton of } M) \cap U_i$ . The projection $p : U_i \to D^2$ locally decreases the area of $X_i$ . It follows, by applying Sard's Theorem, that there exists a number $K$ and a point $y \in D^2$ such that $X_i \cap S^1 \times \{y\}$ has at most $K$ points and $f(\text{2-skeleton of } M)$ is transverse to $S^1 \times \{y\}$ .

We can select a regular neighbourhood $V_i$ of $S^1 \times \{y\}$ and so obtain a handlebody $Y$ in $N$ with the property that $f(\text{2-skeleton of } M)$ meets $Y$ in transverse 2-discs with a uniform bound $L$ on the number of such discs. Thus $f^{-1}(Y)$ consists of handelebodies in $M$ . Now the degree of $f : M \to N$ must equal the degree of $f| : f^{-1}(Y) \to Y$ . As this degree is $1$ , there is a component $X$ of $f^{-1}(Y)$ such that $f| : X \to Y$ has non-zero degree. This degree cannot exceed $L$ . Hence the image of $\pi_1(X)$

in $\pi_1(M)$ is also of index at most $L$. Now $\pi_1(M)$ has only finitely many subgroups of index at most $L$, and for a given handlebody $X$, there are only finitely many homotopy classes of maps $X \to Y$ of degree at most $L$. Also any homomorphism $\pi_1(X) \to \pi_1(N)$ has at most one extension to a homomorphism $\pi_1(M) \to \pi_1(N)$, because of the unique divisibility of hyperbolic elements of $PSL(2, \mathbb{C})$. It follows that there are only finitely many homotopy classes of homotopy equivalences $M \to N$ as required.

## REFERENCES


1. AHLFORS, L., 'Quasi-conformal deformations and mappings in $\mathbb{R}^n$', *Journal d'Analyse Mathématique*, 30 (1976), 74-97.

2. BERS, L., 'Uniformisation by Beltrami equations', *Comm. Pure Appl. Math.*, 14 (1961).

3. SULLIVAN, D., 'Ergodic theory at infinity of an arbitrary discrete group of hyperbolic motions', *Proc. 1978 Stony Brook Conference*, P.U.P.

4. THURSTON, W.P., *The geometry and topology of 3-manifolds*, Lecture notes from Princeton University, 1978-9.

5. THURSTON, W.P., Three dimensional manifolds, Kleinian groups and hyperbolic geometry, Preprint Princeton (1981), 45p.

6. THURSTON, W.P., Hyperbolic structures on 3-manifolds, I: Deformation of acylindrical manifolds, Preprint, Princeton (1981), 57p.

# Sewing-up link exteriors

W. R. BRAKES

Abstract

The purpose of this paper is to describe a construction which produces an abundance of potentially useful examples of three-manifolds, and to demonstrate its applicability to several interesting topics in low-dimensional topology.

## 1. The construction

Let $L$ be a link in $S^3$ of two ordered components $\{K_1, K_2\}$; let $N_1, N_2$ be disjoint, closed, solid torus neighbourhoods of $K_1, K_2$ respectively, with untwisted parametrisations as $S^1 \times D^2$. Then for every $2 \times 2$ integer matrix $A$ with determinant $-1$, $S(L, A)$ denotes the closed orientable manifold formed from the exterior of $L$,

$$E(L) = \overline{S^3 - (N_1 \cup N_2)} ,$$

by identifying the boundary components $\partial N_1$ and $\partial N_2$ via the matrix $A$. Thus each $(x, y)$ (polar coordinates) in

$$S^1 \times \partial D^2 = \partial N_1$$

is identified with $A.\begin{pmatrix} x \\ y \end{pmatrix}$ in $S^1 \times \partial D^2 = \partial N_2$.

## 2. Algebra

For each $i$, let $*_i = (0,0) \in S^1 \times \partial D^2 = \partial N_i$, be the base-point for $\partial N_i$. Select a base-point $*$ in the interior of $E(L)$, and let $u_1, u_2$ be disjoint paths in $E(L)$ from $*$ to $*_1$ and $*_2$ respectively. For $i = 1, 2$, the meridian and longitude $m_i$ and $l_i$ ($0 \times \partial D^2$ and $S^1 \times 0$ respectively in $\partial N_i$) hence define unique elements of $\pi_1(S^3 - L)$, $\mu_i = [u_i m_i u_i^{-1}]$ and $\lambda_i = [u_i l_i u_i^{-1}]$. The passage from $E(L)$ to $S(L, A)$ can be viewed as an appropriate union with torus $\times I$, hence as the addition of one 1-handle, two 2-handles, and one 3-handle. Therefore, if

$$\langle x_1, x_2, \ldots, x_n ; r_1, r_2, \ldots, r_{n-1} \rangle$$

is a presentation of $\pi_1(S^3 - L)$, a presentation of $\pi_1 S(L, A)$, where $A = \begin{pmatrix} \alpha & \beta \\ \gamma & \delta \end{pmatrix}$, is easily found:

$$\langle x_1, \ldots, x_n, y ; r_1, \ldots, r_{n-1}, \lambda_1 y (\lambda_2^{\alpha} \mu_2^{\gamma})^{-1} y^{-1}, \mu_1 y (\lambda_2^{\beta} \mu_2^{\delta})^{-1} y^{-1} \rangle ,$$

where (after identification of $*_1$ with $*_2$ in $S(L, A)$) $y$ is $[u_2 u_1^{-1}]$.

The homology of $S(L, A)$ (all homology is with integer coefficients) can then be quickly calculated:

$$H_1 S(L, A) = \mathbb{Z} \oplus \langle g_1, g_2 ; \alpha n g_1 - (n + \gamma) g_2 = 0, (1 + \beta n) g_1 - \delta g_2 = 0 \rangle ,$$

where $g_1$ and $g_2$ are represented by the cycles $m_1$ and $m_2$, the generator of the $\mathbb{Z}$ summand is represented by the cycle corresponding to $y$, $n$ is the linking number, $lk(K_1, K_2)$, and the usual sign and orientation conventions are adopted (as in [7], for instance). As an example, if $n = 0$, $H_1 S(L, A)$ is $\mathbb{Z} + \mathbb{Z}_{|\gamma|}$.

## 3. Homology handles

From the above, $S(L, A)$ is a homology handle (i.e. $H_* S(L, A) = H_*(S^1 \times S^2)$) if and only if the matrix

$$\begin{pmatrix} \alpha n & -(n + \gamma) \\ 1 + \beta n & -\delta \end{pmatrix}$$

is invertible over the integers, i.e. if and only if

$$\beta n^2 + 2n + \gamma = \pm 1 . \qquad (1)$$

If this condition is satisfied the homology class represented by $u_2 u_1^{-1}$ is a generator for $H_1 S(L, A)$.

In [4] Kawauchi defines an equivalence relation between orientable homology handles (namely $\tilde{H}$-cobordism) and a binary operation ($\circ$ , circle union) on the corresponding equivalence classes, so as to form an abelian group $\Omega(S^1 \times S^2)$ . A basic property of this group is the commutative diagram:

$$\begin{array}{ccc} C^1 & \xrightarrow{e} & \Omega(S^1 \times S^2) \\ & \phi \searrow \quad \swarrow \psi & \\ & G_- & \end{array} \qquad (2)$$

where $C^1$ is the knot concordance group, $G_-$ is the integral matrix cobordism group, $\phi$ and $\psi$ are epimorphisms, and $e$ is the homomorphism induced by longitude surgery (0-surgery).

The sewing-up construction produces a tractable subgroup of $\Omega(S^1 \times S^2)$ , by extending its definition as follows. Let $L^{(r)}$ denote an arbitrary link of two disjoint genus $r$ handlebodies in $S^3$ , $\{H_1 , H_2\}$ , and $f$ any homeomorphism $\partial H_1 \to \partial H_2$ . Then $S(L^{(r)} , f)$ denotes the closed manifold

$$\overline{S^3 - (H_1 \cup H_2)} \,/\, \{\partial H_1 \equiv \partial H_2 \,, x \sim f(x)\} \,.$$

$$\{[M] \in \Omega(S^2 \times S^2) \mid M = S(L^{(r)} , f)\}$$

is a subgroup of $\Omega(S^1 \times S^2)$ , which we'll denote by $S(S^1 \times S^2)$ . The operation of circle union in this subgroup can be readily described. If $L = \{H_1 , H_2\}$ , $L' = \{H_1' , H_2'\}$ are two 'links' (of the generalised type just introduced), *a circle union* of them is defined by placing each in a different hemisphere of $S^3$ (separated by the equator, $S$ ), selecting disjoint arcs $l_1 , l_2$ in $E(L \cup L')$ so that each $l_i$ joins the base points of $\partial H_i$ and $\partial H_i'$ intersecting $S$ in a single point, and defining $L \circ L'$ as

$$\{H_1 \cup N(l_1) \cup H_1' \,, H_2 \cup N(l_2) \cup H_2'\}$$

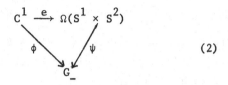

(where each $N(l_i)$ is a regular neighbourhood of $l_i$ in $E(L \cup L')$ ). If $S(L , f)$ and $S(L' , f')$ are two orientable homology handles, $f$ and $f'$ may be assumed to match up the 'ends' of $N(l_1)$ and $N(l_2)$ , and so their restrictions to the remainder of $\partial H_1 \cup \partial H_1'$ extend (non-uniquely) to a homeomorphism

$$f \circ f' : \partial(H_1 \cup N(1_1) \cup H_1') \to \partial(H_2 \cup N(1_2) \cup H_2') .$$

Then $S(L \circ L' , f \circ f')$ is a homology handle. It is easily seen that in $\Omega(S^1 \times S^2)$

$$[S(L \circ L' , f \circ f')] = [S(L , f)] \circ [S(L' , f')]$$

(the right-hand circle as in [4]), so $S(S^1 \times S^2)$ is indeed a subgroup of $\Omega(S^1 \times S^2)$ , and the circle union described above is well-defined up to $\tilde{H}$-cobordism.

## 4. Homology spheres

In [4] Kawauchi demonstrates how it is always possible to do surgery along some curve in any homology handle so as to produce a homology sphere. We can describe this construction explicitly for the examples $S(L , A)$ ( $L$ is a link of circles, though the higher genus case is identical; $A$ still satisfies condition (1) throughout this section). Let $q : E(L) \to S(L , A)$ be the quotient map. Let $w$ be a path $u_1^{-1}u_2$ in $E(L)$ joining $*_1$ to $*_2$ . Let $N_0$ be a regular neighbourhood of $w$ in $E(L)$ so that for $i = 1,2$ , $N_0 \cap \partial N_i$ is a disc $D_i$ , and $q(D_1) = q(D_2)$ . Then $q(N_0)$ is a solid torus. Let $m_0$ be $q(\partial D_1)$ , a meridian of this solid torus, and let $1_0$ be a 'longitude' (not uniquely specified), a curve in the torus $\partial q(N_0)$ intersecting $m_0$ transversely in a single point, and homologous in $q(N_0)$ to its core $q(w)$ . Now if $W$ is a solid torus, the manifold formed from the union

$$\overline{S(L , A) - q(N_0)} \cup W$$

by identifying the boundaries so that the meridian of $W$ is matched with $1_0$ , is a homology sphere. We shall denote this manifold by $SS(L , A ; w , 1_0)$ , the homology sphere got by 'sewing-up' the exterior of $L$ by $A$ , and then 'surgering' along $w$ using $1_0$ .

In the particular case when $L = \{K_1 , K_2\}$ is a split link, $S(L , A)$ is

$$M(K_1 , K_2 ; A) \# S^1 \times S^2 ,$$

where $M(K_1, K_2 ; A)$ is the manifold $E(K_1) \cup_A E(K_2)$ (see Gordon, [3]). If $w$ is restricted to be a path intersecting the sphere that separates $K_1$ from $K_2$ in a single point, then the homology sphere

$$SS(L, A ; w, 1_0)$$

is independent of the particular $w$ and the cho:ce of $1_0$. (To see the lack of dependence on $w$, adapt the usual proof of uniqueness for knot addition; the various choices for $1_0$ can then be shown to be equivalent by twisting $K_1$ (say) whilst fixing $K_2$ ). In fact this homology sphere is just

$$M(K_1, K_2 ; A) .$$

The homology spheres of the type $SS(L, A ; w, 1_0)$ therefore include the Gordon examples as special cases, so in particular include the homology spheres obtained by Dehn's construction (1/q-surgery) on knots in $S^3$. Added complexity may be introduced in two ways:

(i) by choosing a more complicated $w$, e.g. with $L$ as just the unlink and $A = \begin{pmatrix} 0 & 1 \\ 1 & 0 \end{pmatrix}$, one can get the Mazur-Glaser ([6], [2]) type of homology sphere - see Fig.1, where $q(w)$ is the curve $\Gamma$ of [6]; or

(ii) by using a non-splittable link $L$.

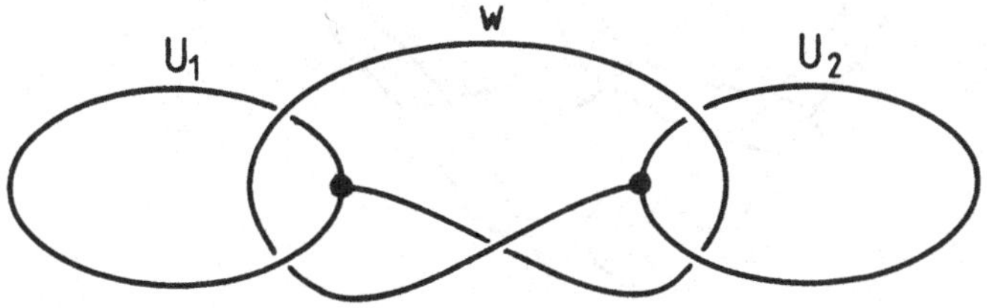


Fig. 1

This SS procedure allows the definition of a collection of concordance invariants for the link $L$, corresponding to every choice of $A$ that satisfies (1) above. These arise from consideration of the core of the added solid torus $W$ in the homology sphere $SS(L, A ; w, 1_0)$. This knot specifies a Seifert matrix and hence an Alexander polynomial (up to the usual equivalences) and signature. These are independent of the choice of $w$ and $1_0$ , since they may be derived from the homomorphism $\phi$ of (2), Section 3 (see [4]).

## 5. Property R

It may happen that the homology sphere $SS(L, A; w, 1_0)$ turns out to be the genuine 3-sphere $S^3$ , in what appear to be unexpected cases. For example, if L is the Whitehead link and J is $\begin{pmatrix} 0 & 1 \\ 1 & 0 \end{pmatrix}$ , then for an appropriate choice of $w$ and $1_0$ ,

$$SS(L, J; w, 1_0) = \overline{S(L, J) - q(N_0)} \cup W \cong S^3 .$$

Under this homeomorphism, the core of W is mapped to the knot shown in Fig.2. Reversing this statement, 0-surgery along the knot of Fig.2 yields the homology handle $S(L, J)$ . Since this manifold is *not* $S^1 \times S^2$ (in fact $\pi_2 S(L, J) = 0$) , the knot of Fig.2 has Property R . This example is of interest because it is a *superslice* (a symmetrically-placed slice of an unknotted 2-sphere), so is not covered by the Kirby/Melvin result [5]. For proofs of these assertions and other examples of this type, see [1].

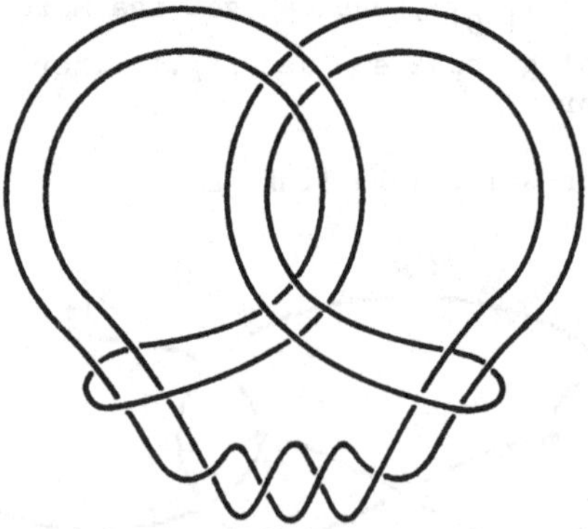

Fig. 2

## 6. Fibred knots

The exterior of a fibred knot K is of the form

$$S \times [0, 1] / \{(x, 0) \sim (h(x), 1)\}$$

for some homeomorphism h of the surface S . Each $\partial S \times t$

$(0 \le t \le 1)$ is a longitude of $K$. So 0-surgery along $K$ produces the closed manifold

$$F \times [0,1] / \{(x,0) \sim (h(x),1)\}$$

where $F$ is a closed surface. This manifold is $S(L^{(r)}, h)$ ($r$ = genus of $F$), where $L^{(r)}$ is the genus $r$ link shown in Fig.3.

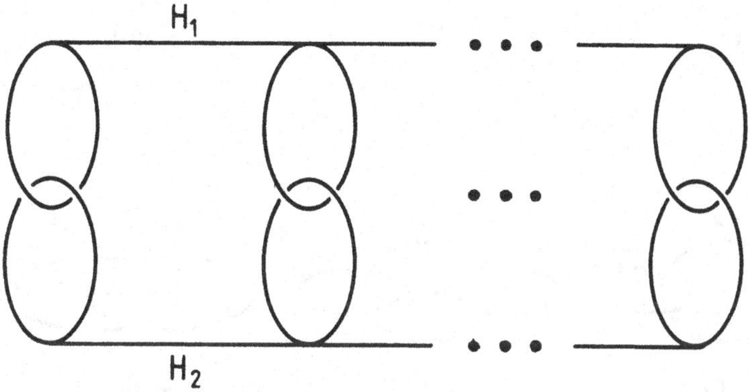


Fig. 3

This process may sometimes be reversed, the sewing-up construction being used to *see* that a knot is fibred. As an example, consider the effect of 0-surgery on the Figure 8 knot, illustrated in the sequence of Fig.4. The first three descriptions are standard. (iii) → (iv) comes from comparing the two views of $S^1 \times S^2$, as the result of 0-surgery along the unknot in $S^3$, and as $S(\{U_1, U_2\}, J)$ where $\{U_1, U_2\}$ is the unlink and $J = \begin{pmatrix} 0 & 1 \\ 1 & 0 \end{pmatrix}$ (see [1]). For (iv) → (v) a spanning disc for $U_2$ is used to cut open the solid torus $E(U_2)$, which is then twisted twice and glued back: this brings the surgery coefficient to the manageable $-1$ at the expense of changing the gluing matrix. (v) → (vi) is the effect of performing the surgery, which links $U_1$ and $U_2$ and also alters their framing. So since 0-surgery along the Figure 8 knot produces

$$T^2 \times [0,1] / \{(x,0) \sim (h(x),1)\}$$

where the homeomorphism $h$ is described by the matrix $\begin{pmatrix} 2 & 1 \\ 1 & 1 \end{pmatrix}$, and the added torus appears as $q(N(w))$ where $w$ is the dotted arc of Fig.4 (iv), (v), (vi), the well-known fibering of the complement of the Figure 8 knot is observed, along with its monodromy.

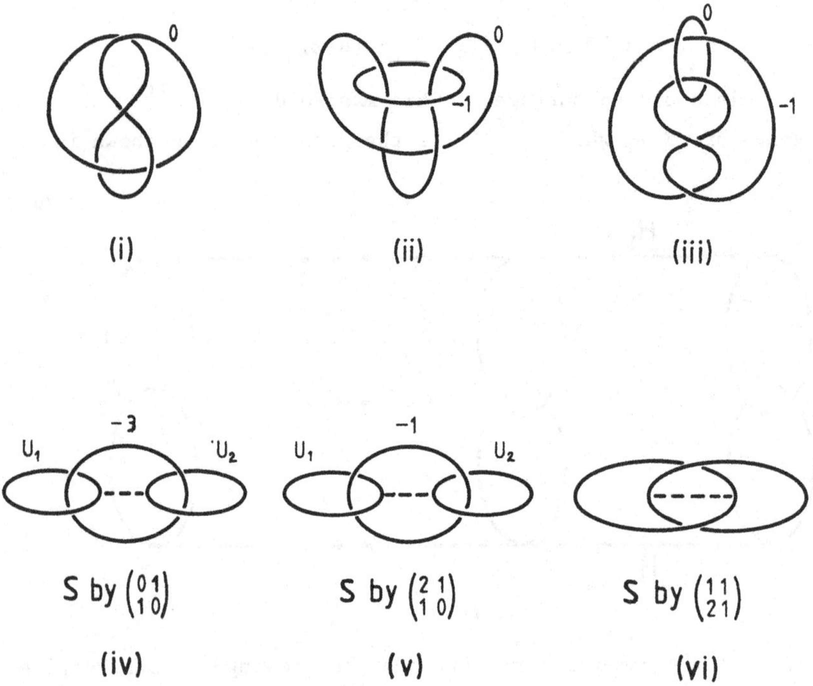


Fig. 4

## 7. Cobordisms

If $L_1$ and $L_2$ are concordant links in $S^3$, there is an obvious way one can use an allowable matrix A to sew-up this concordance, and so produce an interesting cobordism between $S(L_1, A)$ and $S(L_2, A)$. One can use this method to mimic many of the results of [3]. We shall be content here with one example, demonstrating the type of result that is obtainable, the proof being completely analogous to that of Theorem 3 of [3].

THEOREM 7.1 *If* L *is a strongly slice link and* A *is any allowable matrix with* $\gamma \neq \pm 1$, *then* $S(L, A)$ *is homology cobordant to* $S^1 \times S^2$, *by a cobordism with infinite cyclic fundamental group.*

The following theorem partly explains the particular effectiveness of this construction for showing that supersclices have

Property R .

THEOREM 7.2 *Suppose* $L = \{K_1, K_2\}$ *is a link for which* $lk(K_1, K_2) = 0$ *and* $K_2$ *is unknotted. Let* $J = \begin{pmatrix} 0 & 1 \\ 1 & 0 \end{pmatrix}$ . *If* $SS(L, J; w, l_0)$ *is* $S^3$ *for some choice of* $w$ *and* $l_0$ *, so* $S(L, J)$ *is obtainable by 0-surgery along a knot* $k$ *in* $S^3$ *, then this* $k$ *is superslice.*

*Proof.* Let $D$ be an unknotted spanning disc for $K_2$ in the 4-ball $B^4$ . Let $N(D) = D \times B^2$ be a regular neighbourhood of $D$ in $B^4$ so that $\partial D \times B^2 = N(K_2)$ . Let

$$M = \overline{B^4 - N(D)} \cup (S^1 \times B^2 \times I)$$

where $S^1 \times B^2 \times 0$ and $S^1 \times B^2 \times 1$ are identified with

$$D \times \partial B^2 \subset \partial(D \times B^2) = \partial N(D)$$

and $N(K_1)$ respectively. Then $\partial M$ is homeomorphic to $S(L, J)$. Let $C$ be the circle $w \cup (* \times I)$ in $\partial M$ , and let $W = M \cup h^2$ be the result of adding a 2-handle to $M$ along $C$ with some framing. $M$ is obtained by attaching a 'round handle' $S^1 \times B^2 \times I$ to a copy of $S^1 \times B^2 \times I$ , the two ends being attached respectively to a core of $S^1 \times B^2 \times 0$ and a solid torus $N(K_1)$ winding algebraically zero times around $S^1 \times B^2 \times 1$ . Thus, adding a dimension so that this algebraic condition may be reflected in the geometry, $W \times I$ is seen to have a core consisting of a cylinder with its top edge identified with a single point on the lower edge, plus a 'spanning disc' as illustrated in Fig.5. This core is clearly collapsible, so $W \times I$ is homeomorphic to $B^5$ and $W$ is contractible.

Now $\partial W$ is a homology sphere obtained by surgery along $q(w)$ in $S(L, J)$ , so by hypothesis there is a choice of framing for which

$$(\ \partial W\ ,\ \text{belt sphere of}\ H^2\ )$$

is homeomorphic to $(S^3, k)$ . For this choice of framing, $\partial W$ is a 3-sphere bounding $W$ in the 4-sphere $\partial(W \times I)$ , so by the Schoenflies theorem $W$ is a 4-ball. The cocore of $h^2$ is thus a slicing disc for $k$ in the 4-ball.

To see that $k$ is in fact a superslice, we notice that $W \times I$ collapses to (cocore of $h^2$) $\times I$, so

(4-ball, slicing disc for k) $\times I$

is pairwise homeomorphic to an unknotted $(5, 3)$ ball pair.

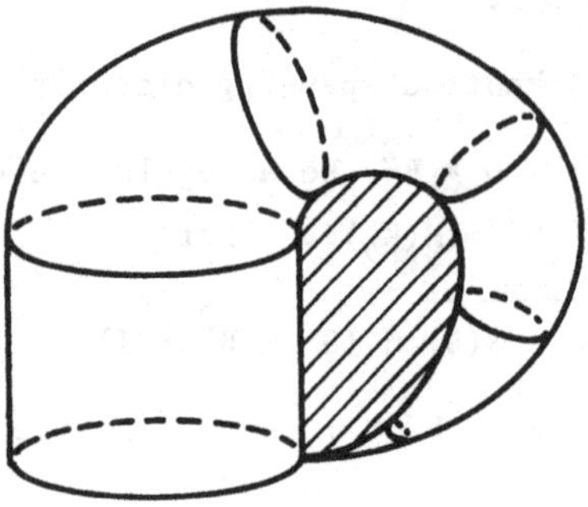

Fig. 5

## 8. Questions

(i) Along which knots in $S^3$ does 0-surgery produce an $S(L^{(r)}, f)$? [Certainly: all fibred knots, all doubles (twisted or otherwise); see [1] for more examples.]

(ii) For which links $L$ and allowable matrices $A$ is $S(L, A)$ obtainable by 0-surgery along some knot in $S^3$? [For instance, is $S(L, A)$ for any *split* link, other than the unlink, so obtainable?]

(iii) Are there any elements of $S(S^1 \times S^2)$ (Section 3) of finite order other than two?

(iv) Are the only elements of $S(S^1 \times S^2)$ of order two those with representatives $S(L^{(r)}, A)$ in which $L^{(r)}$ is negatively amphicheiral?

## REFERENCES

1. W.R. BRAKES, 'Property R and superslices', *Quarterly J. Math.* 31 (1980), 263-281.

2. L.C. GLASER, 'Uncountably many contractible open 4-manifolds', *Topology* 6 (1967), 37-42.

3. C.McA. GORDON, 'Knots, homology spheres, and contractible 4-manifolds', *Topology* 14 (1975), 151-172.

4. A. KAWAUCHI, '$\tilde{H}$-cobordism, I: The groups among three dimensional homology handles', *Osaka J. Math.* 13 (1976), 567-590.

5. R.C. KIRBY, 'Slice knots and Property R', *Invent. Math.* 45 (1978), 57-59.

6. B. MAZUR, 'A note on some contractible 4-manifolds', *Ann. Math.* 73 (1961), 221-228.

7. D. ROLFSEN, *Knots and Links*, Publish or Perish, 1976.

# Periodic transformations in homology 3-spheres and the Rohlin invariant

L. CONTRERAS-CABALLERO

## 1. Introduction

The *Rohlin Invariant* is defined for $\mathbb{Z}_2$-homology 3-spheres $M^3$ (i.e. closed 3-manifolds $M^3$ with $H_*(M^3;\mathbb{Z}_2) = H_*(S^3;\mathbb{Z}_2)$) . Indeed $M^3$ is the boundary of a 4-manifold $W^4$ such that $H_1(W^4;\mathbb{Z}_2) = 0$ and the quadratic intersection form of $H_2(W^4)/(\text{Torsion})$ is even; the Rohlin invariant $\mu M^3$ is by definition

$$\mu M^3 = -\frac{\sigma W^4}{16} \pmod 1 ,$$

where $\sigma W^4$ is the signature of this quadratic form [4] .

It is well-defined, i.e. independent of the manifold $W^4$ , by Rohlin's Theorem [25] .

The existence of a $\mathbb{Z}$-homology 3-sphere $M$ with Rohlin invariant $\frac{1}{2}$ such that $M \# M$ bounds an acyclic manifold would prove the triangulability as simplicial complexes of all manifolds of dimension $\geq 5$ [9] .

A $\mathbb{Z}$-homology 3-sphere with an orientation reversing self-homeomorphism and Rohlin invariant $\frac{1}{2}$ would be a solution to the triangulation problem, since in this case $M \# M = \partial[(M - B^3) \times I]$ .

In this paper, a 3-manifold with an orientation reversing self-homeomorphism is called *symmetric* .

The cyclic and dihedral branched covers of homology 3-spheres with group $D_{2p}$ with branch set an amphichaeiral knot are symmetric. A large class of symmetric closed 3-manifolds are those having a periodic orientation reversing self-homeomorphism since they include the class of symmetric hyperbolic manifolds. Actually, every self-homeomorphism of a hyperbolic 3-manifold is isotopic to an isometry of the manifold [ 19, 22, 29 ] , and these isometries are periodic because the Ricci curvature of a hyperbolic manifold is negative [ 16, 17 ] . The class of hyperbolic manifolds itself is very large because of Thurstons's results [27] .

J.S. Birman [1] , W.C. Hsiang and P. Pao [12] and also D. Galewski and R. Stern [8] separately have proved:

THEOREM 1: *Every homology 3-sphere with an orientation reversing self-diffeomorphism of period 2 has $\mu$-invariant zero.*

I have proved directly:

THEOREM 2: *Every homology 3-sphere with an orientation reversing self-diffeomorphism of period >2 has $\mu$-invariant zero.*

Theorem 1 and Theorem 2 are together expressed in:

THEOREM 3: *Every homology 3-sphere with an orientation reversing periodic self-diffeomorphism has $\mu$-invariant zero.*

My proof runs as follows:

Let $M$ be a homology 3-sphere and $h$ a periodic orientation reversing self-diffeomorphism of $M$ . As the identity preserves orientation, $h^n = 1$ implies $n = 2k$ , for some $k \in \mathbb{N}$ .

If $k$ is odd, we consider $h^k$ , an orientation reversing self-diffeomorphism of $M$ with period 2 . Then Theorem 1 says $\mu M = 0$ .

If $k$ is even, let $2^r m$ be the period of $h$ , $m$ odd . By considering $h^m$ instead of $h$ we may assume $h$ has period $2^r$ , $r > 1$ . The fixed point set of $h$ is $S^o$ by the Lefschetz-Hopf fixed point theorem [4] and Smith theory [7] , since $h$ is orientation reversing and has period $>2$ . Thus the fixed point set of $h^2$ is not empty, and is thus a knot in $M$ by Smith theory. Moreover every non-trivial power of $h^2$ has the same fixed point set by Smith theory, so $M \to M/h^2$ is a cyclic branched cover, branched along a knot $K$ in $M/h^2$ . Also $M/h^2$ is a homology sphere, since $\pi_1(M) \to \pi_1(M/h^2)$ is onto . The orientation reversing homeomorphism $h$ of $M$ projects to an orientation reversing homeomorphism of $M/h^2$ which takes $K$ to itself. Thus K is an amphicaeiral knot in a homology sphere, so $\mu M = 0$ by the Lemma of §3 .

For cyclic covers it is seen from the lemma that if an homology 3-sphere N is a cyclic branched cover of an amphicaeiral knot $K = (M, S^1)$ where M is a homology sphere and either $\mu M = 0$ or the covering degree is even, then $\mu N = 0$ . Thus the class of symmetric homology 3-spheres with μ-invariant zero is enlarged more since the self-diffeomorphisms of these covers may not be periodic.

Even more L. Siebenmann has used my result to show that any symmetric Haken homology 3-sphere has μ-invariant zero [26] .

I am grateful to A.J. Casson for encouraging conversations, and to referees for very helpful comments.

## 2. Preliminary Definitions and Constructions.

Let M be an oriented $\mathbb{Z}$-homology 3-sphere, $K = (M, S^1)$ a knot in M and F a Seifert surface for the knot K . Let us consider $M \times I$ and identify M with $M \times 0$ . We let $F' = (F \times \frac{1}{2}) \cup (K \times [0, \frac{1}{2}])$ , let F(m) be the oriented m cyclic branched cover of $M \times I$ with branch set F' , and let K(m) be the oriented m cyclic branched cover of M with branch set K .

The 4-manifold F(m) is obtained, see [13] , by gluing $F' \times D^2$ to the unbranched cyclic cover of $(M \times I) - (F' \times D^2)$ , namely the pullback

$$\begin{array}{ccc} F(m) - F' \times D^2 & \longrightarrow & S^1 \\ \downarrow & & \downarrow e_m \\ M \times I - F' \times D^2 & \xrightarrow{f} & S^1 \end{array}$$

where $[f] \in H^1(M \times I - F' \times D^2) = [M \times I - F' \times D^2, S^1] = \mathbb{Z}$ is a generator and $e_m(x) = x^m$, $x \in S^1 \subset \mathbb{C}$ .

If f is differentiable and transverse to a point $a \in S^1$ , then $f^{-1}(a)$ is a codimension 1 submanifold such that $F(m) - F' \times D^2$ may be obtained by gluing in a cyclic way m split copies of $M \times I - F' \times D^2$ along $f^{-1}(a)$ . We can extend f to $M \times I$ over $D^2$ and then obtain F(m) by gluing in a cyclic way m split copies of $M \times I$ along $f^{-1}[0, a]$ . We can choose an $f$ such that $f^{-1}[0,a] = F \times [0,\frac{1}{2}]$, where $F \times \{t\} \subset M \times \{t\}$. Then $F(m)$ is obtained by gluing in a cyclic way $m$ copies of $M \times I$ split along $F \times [0,\frac{1}{2}]$.

But if we consider $W = F \times [-\frac{1}{2},\frac{1}{2}] \times 0$, a normal bundle to $F$ in $M \times 0 \subset M \times I$, and define $W^+ = F \times [0,\frac{1}{2}] \times 0$, $W^- = F \times [-\frac{1}{2}, 0] \times 0$ and identify $W^+$ with $W^-$ in $M \times I$ by the relationship $((x,t),0) \sim ((x,-t),0)$, we obtain again $M \times I$ having sent the identification $W^+$ and $W^-$ to $F \times [0,\frac{1}{2}]$. This fact allows us to construct $F(m)$ also, once more by gluing in cyclic order the $m$ copies of $M \times I$ identifying $((x,t),0) \in W^+$ in one copy with $((x,-t),0) \in W^-$ in the other copy.

The manifold $F(m)$ has a $\mathbb{Z}_m$-action induced in $H_2(F(m))$. The signatures of the intersection quadratic form on $H_2(F(m))$ and its restriction to the $\omega$-eigenspace, are written $\sigma F(m)$ and $\sigma_\omega F(m)$ respectively.

## 3. Proof of the Lemma

LEMMA. *Let* $M$, $K$, $K(m)$ *and* $F(m)$ *be as defined previously. If* $K(m)$ *is a* $\mathbb{Z}$*-homology 3-sphere, then:*

i) $\mu K(m) = m\mu M - \sigma F(m)/16 \pmod 1$.

ii) *The signature* $\sigma F(m)$ *is independent of the chosen surface, and is thus just an invariant of the oriented knot* $K = (M,S^1)$. *If the orientation of* $M$ *is reversed,* $\sigma F(m)$ *changes sign, and so is zero for an amphicaeiral knot.*

*Remark.* One can express $\sigma F(m)$ as the sum over all $m$'th roots of unity $\omega$ of the knot signatures $\sigma_\omega K$. These signatures were introduced in various guises for classical knots by Levine, Milnor, Tristram, Cappell and Shaneson, Kauffman and Durfee, Neumann, *et al.* (see [10] and the references given there). Here $\sigma_\omega K$ can be defined either as the signature of the $\omega$-eigenspace of the $\mathbb{Z}_m$-action on $H_2(F(m))$ or as the signature of the hermitian form $(1-\omega)A + (1-\bar{\omega})A$ where $A$ is a Seifert form for $K$. The equivalence of these two definitions is well known, and follows for classical knots from the explicit computation of $H_2(F(m))$ and its intersection form described below.

*Proof.* We verify that $H_1(F(m))$ has no 2-torsion and that the quadratic form of $F(m)$ is even, so that we can write i).

In order to calculate $H_i(F(m))$ we think of $F(m+1)$ constructed from $F(m)$ and one copy of $M \times I$ by splitting $F(m)$

once along $F \times [0,\frac{1}{2}]$ and then gluing on the other copy of $M \times I$. Considering the two pieces:

1) the split $F(m)$ which is homeomorphic to $F(m)$
2) $M \times I$ ,

we obtain the following Mayer-Vietoris sequence:

$$0 \to H_2(F(m)) \to H_2(F(m+1)) \to H_1(F \times I) \to H_1(F(m)) \to H_1(F(m+1)) \to 0 .$$

By induction we conclude that $H_1(F(m)) = 0$ , and that there is an isomorphism $\phi$ from $H_2(F(m))$ to the sum of $m-1$ copies of $H_1(F)$. This isomorphism $\phi$ determines $Q$ , the quadratic intersection form on $H_2(F(m))$ , from the Seifert surface of the knot $K$ , and shows that $Q$ is even, as we now explain.

Let us determine $Q$ : We denote by $F_c$ a Seifert surface in $M$ of a loop $c$ in $F \subset M$ , and by $F_c^i$ , $c^i$ , the copies of $F_c$ , $c$ , in the i'th copy of $M \times 0 \subset F(m)$ . We write $\phi_1, \ldots, \phi_{m-1}$ for the injections $\phi_i\colon H_1(F) \to H_2(F(m))$ defined as $\phi_i[c] = F_c^i \cup -F_c^{i+1}$ . Then the isomorphism $\phi$ is given by

$$\phi = \bigoplus_1^{m-1} \phi_i .$$

An illustrative but not exact picture for $F(2)$ is :

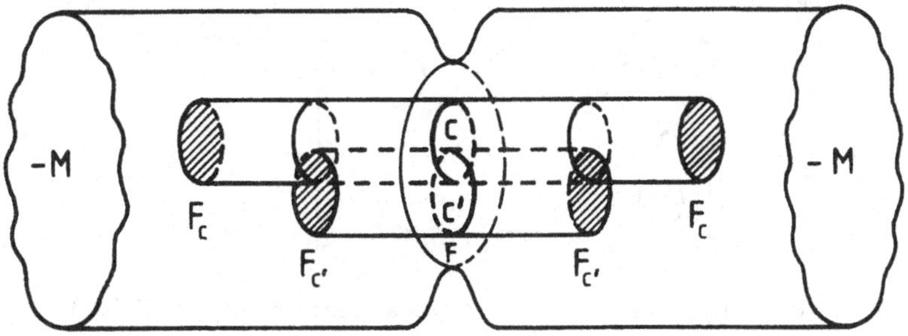

The computation we do is just like the case of classical knots, done by Cappell and Shaneson [2] , Durfee and Kauffman [5] , Gordon [11] , Kauffman [13] , Neumann [23] .

The quadratic intersection form in $F(m)$ is given therefore by

$$\phi_i[c] \cdot \phi_j[c'] = (F_c^i \cup -F_c^{i+1}) \cdot (F_{c'}^j \cup -F_{c'}^{j+1})$$

$$= (F'^{\,i}_{c\times\frac{1}{2}} \cup -F'^{\,i+1}_{c\times-\frac{1}{2}}) \cdot (F_{c'}^j \cup -F_{c'}^{j+1})$$

$$= F'^{\,i}_{c\times\frac{1}{2}} \cdot F_{c'}^j + F'^{\,i}_{c\times\frac{1}{2}} \cdot (-F_{c'}^{j+1})$$

$$+ (-F'^{\,i+1}_{c\times-\frac{1}{2}}) \cdot F_{c'}^j + (-F'^{\,i+1}_{c\times-\frac{1}{2}}) \cdot (-F_{c'}^{j+1})$$

$$= \begin{cases} -\mathrm{lk}(c\times\frac{1}{2}, c') , & \text{if } i = j+1 , \\ \mathrm{lk}(c\times\frac{1}{2}, c') + \mathrm{lk}(c\times-\frac{1}{2}, c') , & \text{if } i = j \\ -\mathrm{lk}(c\times-\frac{1}{2}, c') , & \text{if } i+1 = j , \end{cases}$$

for $[c]$ , $[c'] \in H_1(F)$ .

If $i = j$, $c = c'$ , $\phi_i[c] \cdot \phi_i[c] = 2\mathrm{lk}(c \times \frac{1}{2}, c) \in 2\mathbb{Z}$ shows the quadratic form to be even and so i) is proved.

A basis of $H_1(F)$ of the form $[c_i]$ , $1 \le i \le 2g$ , where the $c_i$ are loops in the orientable surface $F$ , determines a basis $\phi_j[c_i]$ , $1 \le i \le 2g$ , $1 \le j \le m-1$ , of $H_2(F(m))$ . In this basis $Q$ is determined by the Seifert matrix $A$ associated to $K$ by $F$ and expressed by

$$B = \begin{bmatrix} A + A' , & -A & & & \bigcirc \\ -A' , & \ddots & \ddots & & \\ & \ddots & \ddots & \ddots & \\ & & \ddots & \ddots & -A \\ \bigcirc & & & -A' , & A + A' \end{bmatrix}$$

We know that this matrix gives a presentation of $H_1(K(m))$ [10] , so $Q$ is not singular if $K(m)$ is a Q-homology 3-sphere.

In order to prove ii) we consider $F_1$ and $F_2$ , two Seifert surfaces for the knot $K$ , and $F_1(m)$ , $F_2(m)$ the two corresponding covers. Then

$$\partial F_1(m) = \partial F_2(m) = K(m) \coprod_1^m (-M)$$

Identifying the boundaries of $F_1(m)$ and $-F_2(m)$, we obtain a closed four dimensional manifold $H(m)$, which is the m-cyclic branched cover of $M \times S^1$, with branch set $H = F_1' \cup -F_2'$.

The manifold $F_1' \cup -F_2'$ is the boundary of a 3-manifold $G^3$ in $M \times I$ (constructed geometrically from the known relationship between different spanning surfaces of a given knot), and therefore is the boundary of a 3-manifold $G^3$ in $M \times S^1 = \partial M \times D^2$. Now $G^3$ can be pushed in to a proper submanifold $G'$ of $M \times D^2$, such that $H_1(M \times D^2 - G') = \mathbb{Z}$, so that there exists an $m$ cyclic branched cover $G'(m)$ of $M \times D^2$ with branch set $G'$. The boundary of $G'(m)$ is $H(m)$. Thus $\sigma H(m) = 0$ so $\sigma F_1(m) = \sigma F_2(m)$ by Novikov additivity.

Note. After my paper was written I have been informed that A. Kawauchi [14, 15] has obtained in a different way and independently the result of Theorem 2 for $\mathbb{Z}_2$-homology 3-spheres.

## REFERENCES


1. J.S. BIRMAN, 'Orientation reversing involutions on 3-manifolds' Preprint, Columbia University, (1978).

2. S.E. CAPPELL and J.L. SHANESON, 'Branched cyclic coverings', *Knots, groups and 3-manifolds.* Annals of Mathematics Studies 84 (1975), 165-173.

3. L. CONTRERAS-CABALLERO, 'Periodic Transformations in Homology 3-spheres and the Rohlin Invariant.' *Notices of the Amer. Math. Soc.* October 1979, p. A-530.

4. A. DOLD, *Lectures on Algebraic Topology.* Springer-Verlag. Berlin, Heidelberg, New York, 1972.

5. A. DURFEE and L. KAUFFMAN, 'Periodicity of branched cyclic covers', *Math. Ann.* 218 (1975), 157-174.

6. J. EELLS and K.H. KUIPER, 'An invariant for certain smooth manifolds', *Ann. Mat. Pur. Appl.* (4) 60 (1972), 93-110.

7. E.E. FLOYD, 'Periodic maps via Smith theory', *Seminar on Transformation Groups*. Annals of Mathematics Studies 46 (1960), 35-47.

8. D. GALEWSKI and R. STERN, 'Orientation reversing involutions on homology 3-spheres', *Math. Proc. Camb. Phil. Soc.* 85 (1979), 449-451.

9. ------------------------ 'Classification of Simplicial Triangulations of topological manifolds', *Bull. Amer. Math. Soc.* 82 (1976), 916-918.

10. C. McA. GORDON, 'Some aspects of classical knot theory', *Knot theory: Proceedings, Plans-sur-Bex, Switzerland, 1977*. Springer Lecture Notes in Math. 685 (1978), 1-65.

11. -------------- 'Knots, homology spheres and contractible manifolds', *Topology* 14 (1975), 151-172.

12. W.C. HSIANG and P.S. PAO, 'Orientation reversing on homology 3 spheres', *Notices Amer. Math. Soc.* 26, February 1979, p.A-251.

13. L. KAUFFMAN, 'Branched coverings, open books and knot periodicity', *Topology* 13 (1974) 143-160.

14. A. KAWAUCHI, 'On three manifolds admitting orientation reversing involutions', Preprint I.A.S. Princeton, (1979).

15. ----------- 'Vanishing of the Rohlin invariant of some $Z_2$-homology 3-spheres', Preprint I.A.S. Princeton,(1979).

16. S. KOBAYASHI and K. NOMIZU, *Foundations of Differential Geometry*. J. Wiley and Sons, New York, vol. 1, 1963, vol. 2, 1969.

17. S. KOBAYASHI, *Transformation Groups in Differential Geometry*. Springer-Verlag. Berlin, Heidelberg, New York, 1972.

18. J. LEVINE, 'Invariant of knot cobordism', *Inventiones Math.* 8 (1969), 98-110 and 355.

19. A. MARDEN, 'The Geometry of finitely generated Kleinian groups', *Ann. of Math.* 99 (1974), 383-462.

20. J. MILNOR, 'Infinite cyclic coverings', *Conference on the Topology of Manifolds* Prindle, Weber, and Schmidt, Boston, Mass. (1968), 115-133.

21. J. MILNOR and D. HUSEMOLLER, *Symmetric bilinear forms*. Springer-Verlag, Berlin, Heidelberg, New York, 1973.

22. G.D. MOSTOW, 'Strong Rigidity of Locally Symmetric Spaces', *Ann. of Math. Study*, 78, 1976 Princeton Univ. Press.

23. W. NEUMANN, 'Cyclic suspension of knots and periodicity of signature for singularities', *Bull. Amer. Math. Soc.* 80, 977-981, 1974.

24. L.P. NEUWIRTH, *Knot groups*, Annals of Math. Studies 56, Princeton, 1965.

25. W.A. ROHLIN, 'New results in the theory of four dimensional manifolds', *Dokl. Adad. Nauk. SSSR* 84 (1952), 221-224.

26. L. SIEBENMANN, 'On vanishing of the Rohlin invariant and nonfinitely amphicaeiral homology 3-spheres', *Topology Symposium Siegen, 1979*, Springer Lecture Notes in Math. 788 (1980), 172-222.

27. W. THURSTON, 'The Geometry and Topology of 3-manifolds', Preprints, Princeton University, (1978).

28. A.G. TRISTRAM, 'Some cobordism invariants for links', *Proc. Cambridge Phil. Soc.* 66 (1969), 251-264.

29. F. WALDHAUSEN, 'On irreducible 3-manifolds that are sufficiently large', *Ann. of Math.* 87 (1968), 56-58.

# Part 2: Knot theory

# The universal abelian cover of a link

D. COOPER

## 1. Introduction

Given a Seifert surface for a classical knot, there is associated a linking form from which the first homology of the infinite cyclic cover may be obtained. This article considers classical links of two components and shows how to obtain a pair of linking forms from the analogue of a Seifert surface. From these the first homology of the universal abelian $(\mathbb{Z} \oplus \mathbb{Z})$ cover is obtained, thus giving a practical method for calculating the Alexander polynomial. Also obtained is a new signature invariant for links. The method generalises to links of any number of components; however this is not done here.

In this paper, unless otherwise stated, a link will mean a piecewise-linear embedding of two oriented circles in the three sphere $S^3$. The main results are (2.1) and (2.4). The former provides a square matrix presenting the first homology of the cover obtained from the Hurewicz homomorphism of the link complement. The latter gives a signature invariant, obtained from this matrix, which vanishes for strongly slice links. On the way some known results are obtained, namely Torres' conditions on a link polynomial, and a result of Kawauchi and independently Nakagawa on the (reduced) Alexander polynomial of a strongly slice link.

The paper is organised as follows. Section 2 contains the method of obtaining the matrix used in (2.1) and states the main results. The reader not interested in the proofs need read no further. Section 3 contains a proof of (2.1) and also a statement of the Isotopy lemma (3.2). It concludes with a derivation of the Torres conditions. Section 4 is devoted to proving (2.4) and the result on polynomials of slice links.

The material presented here arose out of a study of the method Conway used in [C] to calculate potential functions. A proof of Conway's identities for the Alexander polynomial in a

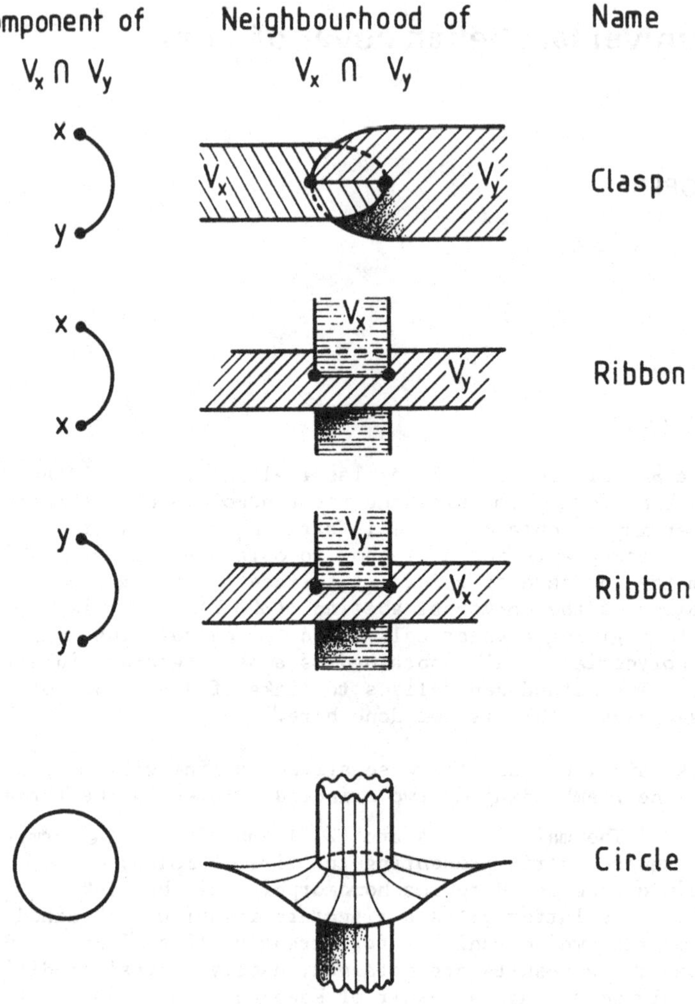


Figure 1.

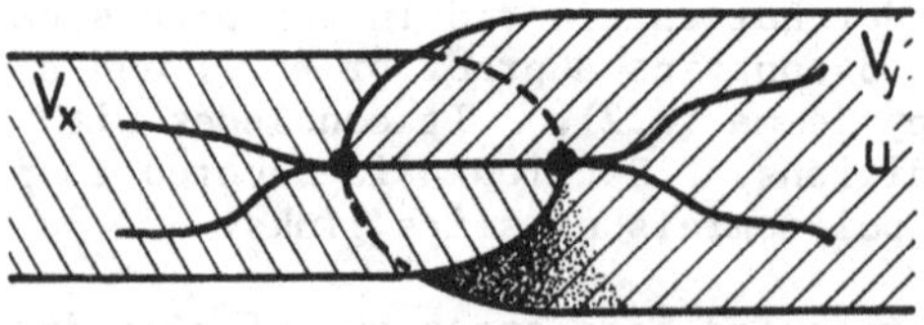


Figure 2. A loop near an intersection

single variable comes from manipulating Seifert surfaces as in [Co], see also Kauffman [K]. The present work enables this proof to be pushed through in the many variable case.

I wish to express my gratitude to Raymond Lickorish and John Conway for their encouragement, and especially to Bill Brakes. This work will form a part of the author's Ph.D. thesis at Warwick University.

## 2. The Algorithm

In this section it is shown how to obtain a pair of matrices from a link. These matrices are used to describe the first homology of the universal abelian cover of the link, to calculate the Alexander polynomial, and to define a new numerical invariant.

Let $V_x$ and $V_y$ be compact pl embedded surfaces in $S^3$ and suppose $V_x$ is disjoint from $V_y$ and that $V_x$ meets $V_y$ transversely. The components of $V_x \cap V_y$ are of three types called *clasp* (or C ), *ribbon* (or R ) and *circle*, see Fig. 1. The 2-complex $S = V_x \cup V_y$ is called a C-*complex* if all intersections are clasps, and R-*complex* if all intersections are ribbon, and an RC-*complex* if ribbon and clasp intersections are allowed. An *orientation* for such a 2-complex is an orientation for each of the component surfaces. The *boundary* of S written $\partial S$ is $(\partial V_x, \partial V_y)$, and the *singularity* of S written $\varepsilon(S)$ is $V_x \cap V_y$.

Given a C-complex S, we define two bilinear forms

$$\alpha, \beta : H_1(S) \times H_1(S) \to \mathbb{Z}$$

as follows. A 1-cycle u in S is called a loop if whenever an ant walking along u meets $\varepsilon(S)$, it does so at an endpoint of some component of $\varepsilon(S)$. That is every component of $u \cap \varepsilon(S)$ has a neighbourhood in S of the form shown in Fig. 2. Given two elements of $H_1(S)$ represent them by loops (this may always be done) u and v and define

$$\alpha([u], [v]) = Lk(u^{--}, v),$$

$$\beta([u], [v]) = Lk(u^{-+}, v),$$

where Lk denotes linking number. $u^{-+}$ is the cycle in $S^3$ obtained by lifting u off S in the negative normal direction off $V_x$ and the positive normal direction off $V_y$. Similarly $u^{--}$ is obtained by using the negative direction for both $V_x$ and

$V_y$. That $u$ is a loop ensures this can be done continuously along $\varepsilon(S)$ where the only difficulty might arise.

Choose a basis $\{\gamma_1,\ldots,\gamma_g\}$ of $H_1(V_x)$ and a basis $\{\gamma_{g+1},\ldots,\gamma_{g+h}\}$ of $H_1(V_y)$ and, identifying via inclusion, extend to a basis $\{\gamma_1,\ldots,\gamma_{g+h+k}\}$ of $H_1(S)$. Let $A$ and $B$ be the integral matrices of $\alpha$ and $\beta$ respectively using this basis.

Suppose now that $L$ is a link of two oriented circles pl embedded in $S^3$ called $L_x$ and $L_y$; we write $L = (L_x, L_y)$. A C-*complex* for $L$ is a connected orented C-complex $S$ such that $\partial S = L$. (Lemma (3.1) says that any pair of Seifert surfaces for $L$ may be deformed into a C-complex for $L$). The Hurewicz homomorphism $\pi_1(S^3 - L) \to H_1(S^3 - L)$ induces a regular cover $\tilde{X}$ of $S^3 - L$, the universal abelian cover, and $H_1(\tilde{X})$ has a natural $\Lambda$-module structure where $\Lambda = \mathbb{Z}[x,y,x^{-1},y^{-1}]$ is the integral group-ring in two variables $x$ and $y$ representing the deck transformations induced by the meridians of $L_x$ and $L_y$. Define a $(g+h+k) \times (g+h+k)$ matrix $J$ over the field of fractions of $\Lambda$ by

$$\begin{aligned} J_{r,s} &= 0 && 1 \le r \neq s \le g+h+k \\ J_{r,r} &= (y-1)^{-1} && r \le g \\ &= (x-1)^{-1} && g+1 \le r \le g+h \\ &= 1 && g+h+1 \le r. \end{aligned}$$

THEOREM 2.1. *If* $S$ *is connected,* $H_1(\tilde{X})$ *is presented as a* $\Lambda$-*module by the matrix* $J(xyA - A^T - xB - yB^T)$. *In particular this matrix has entries in* $\Lambda$.

J. Bailey has obtained a presentation for $H_1(\tilde{X})$ by different means, see [B].

COROLLARY 2.2. *The Alexander polynomial of* $L$ *is*

$$\Delta(x, y) = (y-1)^{-g}(x-1)^{-h} \det(xyA + A^T - xB - yB^T)$$

*where* $\quad g = 2 \times \mathrm{genus}(V_x) \qquad h = 2 \times \mathrm{genus}(V_y)$.

The Alexander polynomial as given in (2.2) may vanish, in

which case the determinant of a presentation mstrix for the torsion submodule of $H_1(\tilde{X})$ I call the *reduced Alexander polynomial* written $\Delta_{red}(x, y)$ .

A link is *strongly slice* if its components bound disjoint locally flat discs properly embedded in the 4-ball.

THEOREM 2.3. [Kaw], [N]. *If* L *is strongly slice then* $\Delta(x, y) = 0$ *and* $\Delta_{red}(x, y) = F(x, y)\, F(x^{-1}, y^{-1})$ *for some* $F(x, y) \in \mathbb{Z}[x, y]$ .

Let $\omega_1, \omega_2$ be complex numbers of modulus 1 and let M be the hermitian matrix $(1 + \bar{\omega}_1\bar{\omega}_2)(\omega_1\omega_2 A + A^T - \omega_1 B - \omega_2 B^T)$ .

Define
$$\sigma(\omega_1, \omega_2) = \text{signature } (M)$$
$$n(\omega_1, \omega_2) = \text{nullity } (M)$$

THEOREM 2.4. *(i)* $\sigma$ *and* n *are invariants of* L *provided* $(1 + \bar{\omega}_1\bar{\omega}_2) \neq 0$ *and* $\omega_1, \omega_2 \neq 1$ . *(ii) If* L *is strongly slice then* $\Delta_{red}(\omega_1, \omega_2) \neq 0 \Longrightarrow \sigma(\omega_1, \omega_2) = 0$ .

Conway has suggested that it is more natural to consider
$$\text{signature } (\omega_1\omega_2 A + \bar{\omega}_1\bar{\omega}_2 A^T - \omega_1\bar{\omega}_2 B - \bar{\omega}_1\omega_2 B^T)$$
in place of the above. This has the advantage of removing the jump in $\sigma$ at $1 + \bar{\omega}_1\bar{\omega}_2 = 0$ at the expense of replacing the connection with the Alexander polynomial by his potential function.

## 3. Homology of the cover

First it is proved that any pair of Seifert surfaces for a link may be deformed into a C-complex. The idea in the proof can be used to prove the isotopy lemma (3.2) for C-complexes which gives a pair of moves by means of which two C-complexes with isotopic component surfaces may be transformed into each other. This result is used in Section 4 to provide a proof of invariance of the signature introduced in Section 2. Theorem 2.1 is proved, and finally a new derivation of the Torres conditions is given.

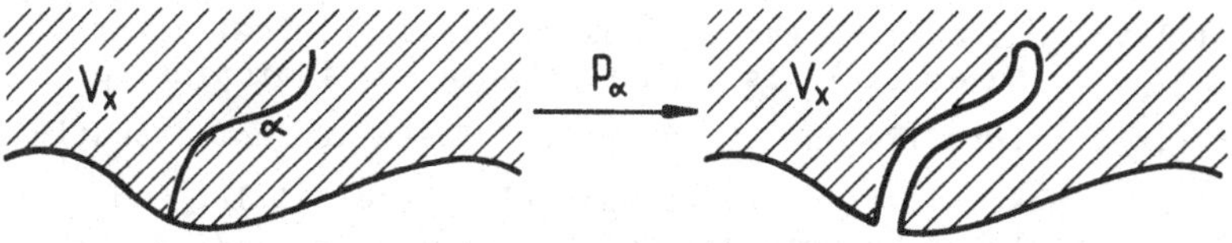


Figure 3. Pushing along an arc.

DEFINITION. Given a surface $V$ with boundary, and an arc $\alpha : [0,1] \to V$ with $\alpha(0)$ the only point on $\partial V$, a *push* along $\alpha$ is an embedding $p_\alpha : V \to V$ defined by choosing two regular neighbourhoods of $\alpha$, $N_1$ and $N_2$, meeting $\partial V$ regularly, with $N_1 \subset \operatorname{Int} N_2$. Then $p_\alpha | (V - \operatorname{Int} N_2) =$ identity and $p_\alpha$ maps $N_2$ homeomorphically onto $N_2 - \operatorname{Int} N_1$, see Fig. 3. Given a pair of Seifert surfaces $V_x$ and $V_y$ for a link, a push along an arc $\alpha$ in $V_x$ is allowed only if $N_2 \cap \partial V_y = \emptyset$. Similarly for a push in $V_y$. That is you are not allowed to push one boundary component through the other.

LEMMA 3.1. *Any pair of Seifert surfaces for a link may be isotoped keeping their boundaries fixed to give a C-complex.*

*Proof.* First make the surfaces transverse, and then remove an outermost-on-$V_x$ circle component of $V_x \cap V_y$ by pushing in along an arc in $V_x$ going from $\partial V_x$ to that circle. This transforms the circle into a ribbon intersection. Continue in this way until all circles have been removed; note that this process does not introduce new circles. Next remove the ribbon intersections, in any order, by pushing along an arc from the boundary of one of the surfaces to the ribbon intersection to replace it by two clasps. The resulting isotopy has moved the link, but only by an ambient isotopy.

ISOTOPY LEMMA 3.2. *Suppose* $S = V_x \cup V_y$ *and* $S' = V'_x \cup V'_y$ *are C-complexes for a link and that* $V_x$ *is isotopic* rel $\partial V_x$ *to* $V'_x$ *and* $V_y$ *is isotopic* rel $\partial V_y$ *to* $V'_y$. *Then* $S$ *may be transformed into* $S'$ *by the following moves and their inverses:*

*(I0)* *Isotope* $S$

*(I1)* *Add a ribbon intersection between* $V_x$ *and* $V_y$, *see Fig.4*

*(I2)* *Push in along an arc in* $V_x$ *or* $V_y$.

The idea of the proof is to make the isotopies of $V_x$ and $V_y$ critical level and examine the various possible critical points of $V_x \cap V_y$. Move (I2) is used to change the isotopies so that critical points lie on $\partial V_x \cup \partial V_y$. There are now essentially only 3 possibilities other than (I1) and (I2), and these may be replaced by combinations of (I1) and (I2). The details appear in [Co].

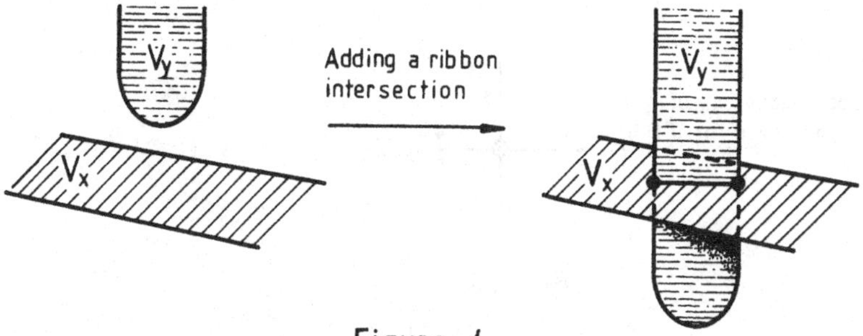


Figure 4.

It is well known that any two Seifert surfaces for a knot are equivalent under adding handles and isotopy. Combining this with the above gives the equivalence relation between C-complexes with the same boundary.

We turn now to homology. Let $S$ be an oriented connected C-complex, then a neighbourhood of a clasp has a cross-section as in Fig. 5. Cut each clasp as shown in Fig. 6(i) to yield an oriented surface $V_{--}$ homotopy equivalent to $S$ by inclusion. Let $V_{--} \times [-1, 1]$ be a bicollar of $V_{--}$ with the $+1$ side as in Fig. 6(i), and let $j$ be the inclusion $V_{--} \longrightarrow V_{--} \times 1 \longrightarrow S^3$. Define a homomorphism $i_{--}$ by requiring the following diagram to commute

$$\begin{array}{ccc} H_1(S) & \xrightarrow{\quad i_{--} \quad} & H_1(S^3 - S) \\ (\mathrm{incl})_* \uparrow \cong & & \cong \downarrow (\mathrm{incl})_* \\ H_1(V_{--}) & \xrightarrow{\quad j_* \quad} & H_1(S^3 - V_{--}) . \end{array}$$

Similarly define $i_{-+}$, $i_{+-}$ and $i_{++}$ by using Figs. 6(ii), (iii) and (iv) respectively. The linking forms

$$\alpha , \beta : H_1(S) \times H_1(S) \to \mathbb{Z}$$

are defined by

$$\alpha(e_1 , e_2) = \mathrm{Lk}\ (i_{--}e_1 , e_2) ,$$
$$\beta(e_1 , e_2) = \mathrm{Lk}\ (i_{-+}e_1 , e_2) .$$

Suppose now that $S$ is a C-complex for a link $L$ and let $p : \tilde{X} \to S^3 - L$ be the universal abelian cover. Then $p^{-1}(S)$ separates $X$ into components homeomorphic to $S^3 - S$. If $S$

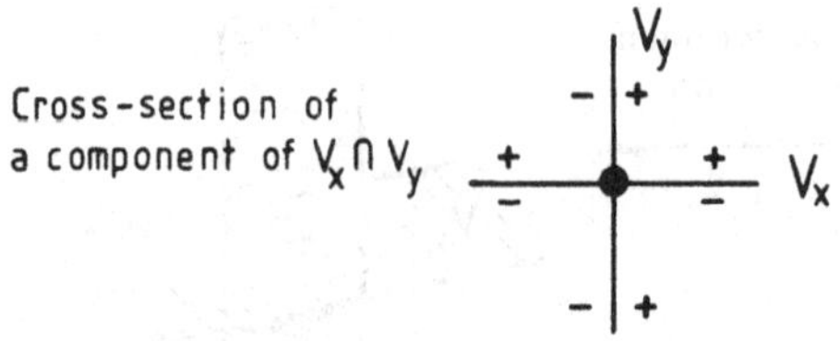


Figure 5.

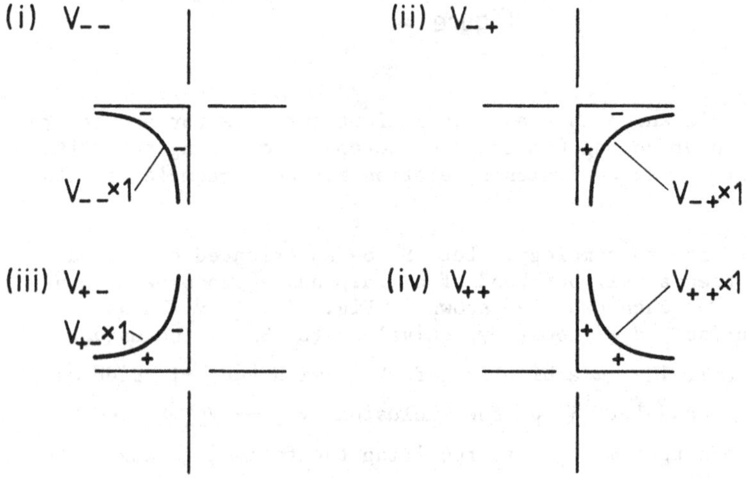


Figure 6.

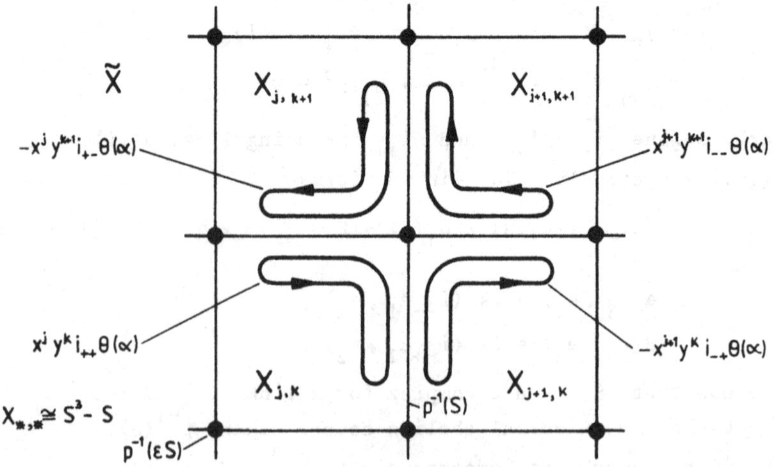


Figure 7.

is connected, $p^{-1}(S)$ is connected, and so $H_1(\tilde{X})$ is generated, as a $\Lambda$-module, by (lifts of) $H_1(S^3 - S)$. By duality $H_1(S^3 - S) \cong H_1(S)$ and by Mayer-Vietoris

$$H_1(S) \cong H_1(V_x) \oplus H_1(V_y) \oplus \hat{H}_0(V_x \cap V_y)$$

where $\hat{H}_*$ is reduced homology. Let $\theta$ be the isomorphism

$$\theta : H_1(V_x) \oplus H_1(V_y) \oplus \hat{H}_0(V_x \cap V_y) \to H_1(S^3 - S) .$$

Regarding $H_1(\tilde{X})$ as a $\Lambda$-module generated by $H_1(S^3 - S)$ a complete set of relations is:

For $\alpha \in H_1(V_x)$ $\quad i_{++}\theta(\alpha) = x.i_{--}\theta(\alpha)$

For $\alpha \in H_1(V_y)$ $\quad i_{++}\theta(\alpha) = y.i_{--}\theta(\alpha)$

For $\alpha \in \hat{H}_0(V_x \cap V_y)$ $\quad i_{++}\theta(\alpha) = x.i_{-+}\theta(\alpha) + y.i_{+-}\theta(\alpha) - xy.i_{--}\theta(\alpha)$

The proof of this is by a double application of the Mayer-Vietoris sequence, see [Co]. The only relations it is hard to visualise is the third set which is suggested by Fig. 7.

It is obvious that for $\alpha \in H_1(V_x)$

$$i_{--}\theta(\alpha) = i_{-+}\theta(\alpha) \quad \text{and} \quad i_{+-}\theta(\alpha) = i_{++}\theta(\alpha)$$

Similarly for $\alpha \in H_1(V_y)$

$$i_{--}\theta(\alpha) = i_{+-}\theta(\alpha) \quad \text{and} \quad i_{-+}\theta(\alpha) = i_{++}\theta(\alpha)$$

The relations may thus be rewritten

For $\alpha \in H_1(V_x)$ $\quad (y-1)^{-1}(xy.i_{--} + i_{++} - x.i_{-+} - y.i_{+-})\theta(\alpha) = 0$

For $\alpha \in H_1(V_y)$ $\quad (x-1)^{-1}(xy.i_{--} + i_{++} - x.i_{-+} - y.i_{+-})\theta(\alpha) = 0$

For $\alpha \in \hat{H}_0(V_x \cap V_y)$ $\quad (xy.i_{--} + i_{++} - x.i_{-+} - y.i_{+-})\theta(\alpha) = 0$

Using the basis of $H_1(S)$ given in Section 2 proves Theorem 2.1. □

THEOREM 2.1. (Torres). *The Alexander polynomial of a link* L *of two components satisfies*

*(i)* $\Delta(x, y) \doteq \Delta(x^{-1}, y^{-1})$

*(ii)* $\Delta(x, 1) \doteq \Delta(x) . (1 - x^{|\ell|})/(1 - x)$

*where* $\doteq$ *denotes equal up to multiplication by* $\pm x^r y^s$ *, and* $\ell$ *is the linking number of the two components.*

Torres' proof [Tor] made use of Fox's calculus. Here is a proof using theorem 2.1; (i) is immediate. For (ii), using the basis of $H_1(S)$ given in Section 2, the linking matrices $A, B$ have the form

$$A = \begin{array}{c} H_1(V_x) \\ H_1(V_y) \\ H_1(\varepsilon S) \end{array} \overset{\begin{array}{ccc} H_1(V_x) & H_1(V_y) & H_1(\varepsilon S) \end{array}}{\begin{bmatrix} C & D & E \\ D^T & F & G \\ H & J & K \end{bmatrix}} \qquad B = \begin{bmatrix} C & D & E \\ D^T & F^T & J^T \\ H & G^T & L \end{bmatrix}$$

Restrict the basis of $H_1(S)$ by requiring that the loops representing the basis of $\hat{H}_0(\varepsilon S)$ are disjoint from those representing the basis of $H_1(V_y)$. This gives $G = J^T$, and

$$(x, 1) = \det \begin{bmatrix} xC-C^T & (x-1)D & xE-H^T \\ 0 & F - F^T & 0 \\ 0 & 0 & x(K-L) + (K-L)^T \end{bmatrix}$$

$$= \det(xC - C^T) \det(F - F^T) \det(xM + M^T)$$

where $M = K - L$.

Now $C$ is a Seifert matrix for the x-component so $\det(xC - C^T)$ = Alexander polynomial for x-component. $F$ is a Seifert matrix for the y-component so $\det(F - F^T) = 1$. Finally it is shown below that $\det(xM + M^T)$ depends only on the linking number of the two components, and evaluating for a simple link gives $(1 - x^{|\ell|}) / (1 - x)$.

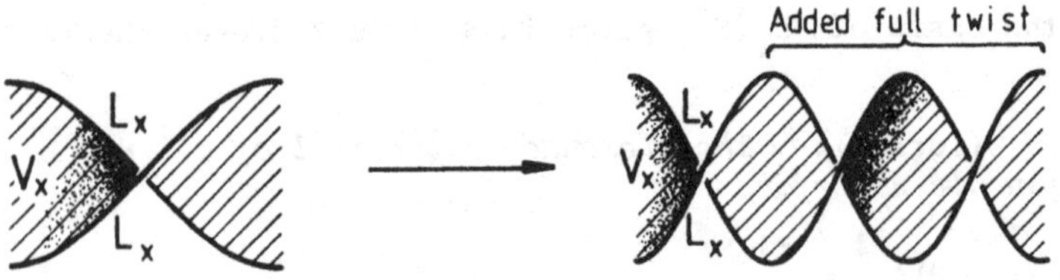


Figure 8.

It is well known that any knot may be changed into the unknot by changing crossovers. This is easily extended to: any link may be changed into the simple link of the same linking number by changing crossovers at which both strings belong to the same component. Let L' be the link L with a single such crossover changed. Choose a C-complex S for L ; then a C-complex S' for L' is obtained by adding a full twist to S next to the changed crossover, see Fig. 8. The matrix M is the matrix of the form $(\alpha - \beta)$ restricted. Adding a twist to S changes $\alpha$ and $\beta$ by adding to each a (symmetric) form $\gamma$ . Thus $(\alpha - \beta)$ is unchanged, and so $\det(xM + M^T)$ is unchanged, completing the proof. □

The Torres conditions are known to characterise link polynomials when the linking number of the two components is 0 or 1 , see [B] , [L]. On the other hand, Hillman has shown in [H] that they are not sufficient for linking number 6.

## 4. Cobordism Invariance of Signature

First, linking forms are introduced for an RC-complex and then the Isotopy lemma is used to show signature is a link invariant. Next is a well known result on the rank as a $\Lambda$-module of $H_1(\tilde{X})$ . This is used in the proofs of theorems 2.3 and 2.4 for ribbon links. Finally the proofs are extended to slice links. The method of proof of 2.4 is similar to that used by Murasugi in [M] and Tristram in [T].

DEFINITION. Let F be an RC-complex and choose a ribbon intersection r in F . Remove from F a small disc centred on the mid point of r and lying in the component surface of F in which the endpoints of r are interior points, see Fig. 9. Let S be a C-complex obtained from F by the above construction at each ribbon intersection, then S is said to be *obtained by puncturing* F . The linking forms $\alpha$ , $\beta$ for S are *associated to* F .

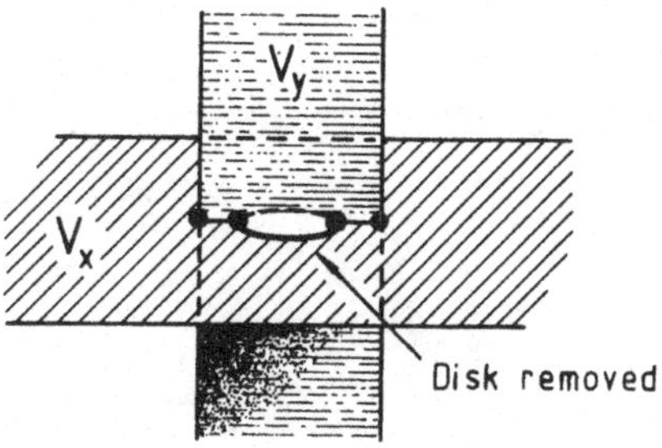


Figure 9.

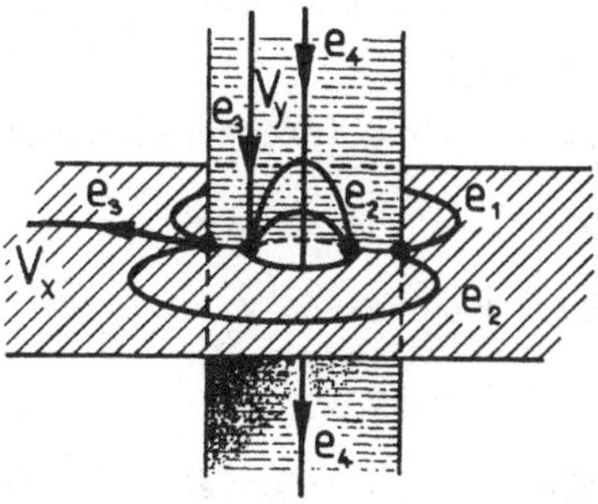


Figure 10.

Let $F$ be an RC-complex and $F_1$ an RC-complex obtained from $F$ by pushing in along an arc $\alpha$ to convert some ribbon intersection $r$ into two clasps. Let $S$ be a C-complex obtained by puncturing $F$, and $S_1$ a C-complex obtained by puncturing $F_1$; we may suppose $S_1 = S \cap F_1$. Choose a neighbourhood $U$ of $r$ in $S$ of the form shown in Fig. 10. Pick loops $\{e_1, \ldots, e_n\}$ representing a basis of $H_1(S)$ such that $e_i$ misses $U$ for $i > 4$ and $e_i \cap U$ is as shown in Fig. 10 for $i \leq 4$. The loops $\{e_2, \ldots, e_n\}$ represent a basis of $H_1(S_1)$. The matrix $(1 + \bar{\omega}_1 \cdot \bar{\omega}_2)(\omega_1\omega_2 A + A^T - \omega_1 B - \omega_2 B^T)$ for $S$ using this basis is

$$Q = \left[\begin{array}{cc|ccc} 0 & 0 & \theta & \phi & -0- \\ 0 & 0 & -\bar{\omega}_1\theta & -\bar{\omega}_1\phi & -0- \\ \hline \bar{\theta} & -\omega_1\bar{\theta} & & & \\ \bar{\phi} & -\omega_1\bar{\phi} & & * & \\ | & | & & & \\ 0 & 0 & & & \\ | & | & & & \end{array}\right] \qquad \begin{array}{l} \theta = \omega_1 + \bar{\omega}_2 \\ \\ \phi = (1 - \omega_2)\,\theta \end{array}$$

Let $Q_1$ be the matrix obtained from $Q$ by omitting the first row and column, so it is the corresponding matrix for $S_1$. Then

$$\text{Signature }(Q) = \text{Signature }(Q_1)$$

$$\text{nullity }(Q) = \text{nullity }(Q_1) + 1 .$$

By the remark after the Isotopy lemma, in order to complete the proof of theorem 2.4(i) it suffices to calculate the effect of:

(I1) Add a ribbon intersection between $V_x$ and $V_y$

(I2) Push in along an arc

($H_x$) Add a handle to $V_x$

($H_y$) Add a handle to $V_y$

The above calculation shows (I2) has no effect on signature.

If $P$ is an hermitian matrix and

$$P_1 = \left[\begin{array}{c|cc} P & v & o \\ \hline v^\dagger & u & w \\ -o- & \bar{w} & o \end{array}\right]$$

where $w$ is a non-zero complex number, $u$ a real, $v$ a complex column vector and $v^\dagger$ its conjugate transpose, then $P_1$ is called an *elementary enlargement* of $P$. The effect of (I1), (Hx) and (Hy) on $Q$ is an elementary enlargement with for

(I1) $w = 1 + \bar{\omega}_1 . \bar{\omega}_2$ or $\omega_1 + \omega_2$

(Hx) $w = |1 + \bar{\omega}_1 . \bar{\omega}_2|^2 (1 - \bar{\omega}_2)$

(Hy) $w = |1 + \bar{\omega}_1 . \bar{\omega}_2|^2 (1 - \bar{\omega}_1)$

Thus signature and nullity are independent of the C-complex used for a link, provided $w \neq 0$ in the above, thus proving 2.4(i). □

LEMMA 4.1. *Let* $L = (L_x , L_y)$ *be a link of two components. Let* $M$ *be the* $\Lambda$*-module which is the first homology of the universal abelian cover. Then rank* $(M) = 0$ *or* $1$ .

*Proof*. If $\Delta_L(x , y) \neq 0$ then $M$ is a torsion module so rank $M = 0$ . Otherwise notice that $\Delta_L(1 , 1) = Lk(L_x , L_y) = 0$ . Choose a C-complex $S$ for $L$ and let $S_1$ be obtained from $S$ by removing one clasp so that $S_1$ is a C-complex for a link $L_1$ with $\Delta_{L_1}(1 , 1) = \pm 1$ . Thus the module for $L_1$ , $M_1$ is a torsion module. Putting back the clasp corresponds to adding a single row and column to a presentation matrix for $M_1$ giving a presentation matrix for $M$ . This latter matrix thus has nullity (equal to rank $M$ ) at most $1$ . □

DEFINITION. Let $D_x$ and $D_y$ be 2-discs immersed in $S^3$ without triple points, with $\partial D_x \cap \partial D_y = \emptyset$ , and so that the only intersections self and mutual are of ribbon type. Then $(\partial D_x , \partial D_y)$ is called a *ribbon link*.

Suppose that $L$ is a ribbon link, the immersed discs $D_x$ and $D_y$ may be cut along their self intersections to give

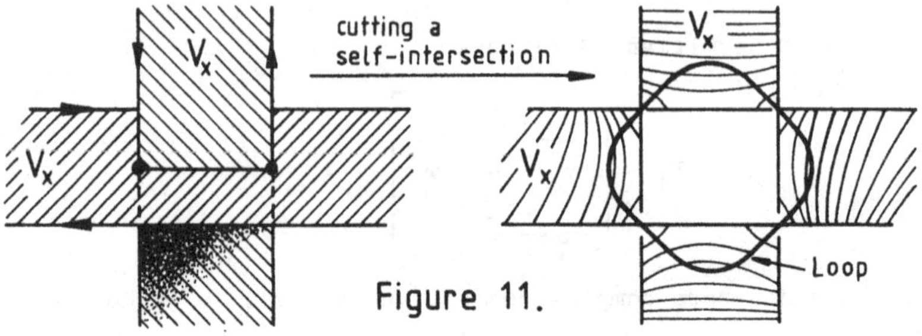


Figure 11.

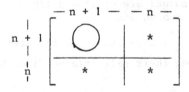

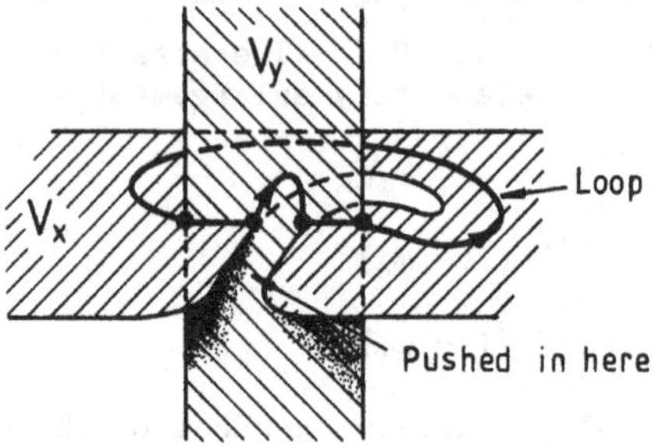


Figure 12.

orientable surfaces $V_x$ and $V_y$ with $L = (\partial V_x, \partial V_y)$, see Fig. 11. $F = V_x \cup V_y$ is an R-complex for $L$; push in to get a C-complex $S$. Pick loops in $S$ representing an ordered basis of $H_1(S)$ as follows:

(1) For each self intersection of $D_x$ pick a loop going around that intersection-cut-open in $V_x$, see Fig. 11.

(2) Do the same for $V_y$.

(3) For each ribbon intersection of $F$ pick a loop in $S$ going through the two resulting clasps in $S$, see Fig. 12.

(4) Complete the basis by picking a further $n$ loops.

The dimension of $H_1(S)$ is easily seen to be $2n + 1$. The matrices $A$ and $B$ using this basis have the shape

$$\begin{array}{c} \\ n+1 \\ n \end{array} \begin{array}{c} \overbrace{\quad\;}^{n+1} \quad \overbrace{\quad}^{n} \\ \left[\begin{array}{c|c} \bigcirc & * \\ \hline * & * \end{array}\right] \end{array}$$

Thus $\Delta(x, y) = 0$, and by lemma 4.1

$$\text{nullity } (xyA + A^T - xB - yB^T) = 1 .$$

Using the method of proof for 4.1 we may assume the first row and column of A and B are zero. Then it follows that

$$\Delta_{red}(x, y) = F(x, y) . F(x^{-1}, y^{-1})$$

for some $F(x, y) \in \mathbb{Z}[x, y]$, and also that

$$\Delta_{red}(\omega_1, \omega_2) \neq 0 \Rightarrow \sigma(\omega_1, \omega_2) = 0 .$$

This proves 2.3 and 3.4(i) for L a ribbon link.

DEFINITION. $L = (L_x, L_y)$ is a link in $S^3$ and $U_1, \ldots, U_n$ are unknots in $S^3$ separated from L and from each other by 2-spheres. $b_1, \ldots, b_n$ are bands, that is disjoint embeddings $I \times I \hookrightarrow S^3$ with $b_i(0 \times I) \subset U_i$ and $b_i(1 \times I) \subset L$. The link $L' = (L'_x, L'_y)$ defined by

$$L' = L \cup \bigcup U_i - \bigcup b_i(\partial I \times I) \cup \bigcup b_i(I \times \partial I)$$

is said to be *obtained from* L *by band-summing an unlink.*

It is well known that a knot is slice if and only if it may be made into a ribbon knot by band-summing an unlink, see for example Tristram [T]. It follows that a link is strongly slice if and only if it may be made into a ribbon link by band-summing an unlink.

Let $L = (L_x, L_y)$ be a link and $L' = (L'_x, L'_y)$ be obtained from L by band-summing an unlink. Choose a C-complex S for L and discs $D_1, \ldots, D_n$ spanning the unlink $U_1, \ldots, U_n$ disjoint from each other and from S, and such that $S \cup D_i$ is transverse to the bands used in band-summing $\bigcup b_i$. From the RC-complex $S \cup \bigcup D_i \cup \bigcup b_i$ form a C-complex S' for L' in a manner similar to that used for making the C-complex for a ribbon link above.

Let A, B be the linking matrices for S. Using the same methods as in the proof for ribbon links above we obtain linking matrices $A_1, B_1$ for S' of the shape

$$A_1 = \begin{bmatrix} 0 & C & 0 \\ D & * & * \\ 0 & * & A \end{bmatrix} \quad B_1 = \begin{bmatrix} 0 & E & 0 \\ F & * & * \\ 0 & * & B \end{bmatrix}$$

The matrices $C, D, E, F$ being square and of the same size. It follows that

$$\Delta_{L'}(x, y) = F(x, y) \cdot F(x^{-1}, y^{-1}) \cdot \Delta_L(x, y)$$

where $\quad F(x, y) = \det(xyC + D' - xE - yF')$ .

Also $\quad F(\omega_1, \omega_2) \neq 0 \Rightarrow \sigma_{L'}(\omega_1, \omega_2) = \sigma_L(\omega_1, \omega_2)$

$$n_{L'}(\omega_1, \omega_2) = n_L(\omega_1, \omega_2)$$

This completes the proof of 2.3 and 2.4(ii).

## REFERENCES


[B] BAILEY, J.H., 'Alexander Invariants of Links', Ph.D. thesis, University of British Columbia, 1977.

[C] CONWAY, J.H., 'An enumeration of knots and links, and some of their algebraic properties. Computational problems in Abstract Algebra', *Pergamon Press*, Oxford and New York (1969) 329-358.

[Co] COOPER, D., Ph.D. thesis, University of Warwick. In preparation.

[H] HILLMAN, J.A., 'The Torres Conditions are Insufficient', preprint.

[K] KAUFFMAN, L., 'The Conway Polynomial', preprint.

[Kaw] KAWAUCHI, A., 'On the Alexander Polynomials of cobordant Links', *Osaka J. Math.* 15 (1978) 151-159.

[L] LEVINE, J.P., 'A Method for generating link polynomials', *Amer. J. Math.* 89 (1976) 69-84.

[M] MURASUGI, K., 'On a certain numerical invariant of link types', *Trans. Amer. Math. Soc.* 117 (1965) 387-422.

[N] NAKAGAWA, Y., 'On the Alexander polynomial of Slice links', *Osaka J. Math.* 15 (1978) 161-182.

[Tor] TORRES, G., 'On the Alexander polynomial', *Ann. of Math.* 57 (1953) 57-89.

[T] TRISTRAM, A.G., 'Some Cobordism Invariants for links', *Proc. Cambridge Philos. Soc.* 66 (1969) 251-264.

# Levine's theorem – a remark

L. H. KAUFFMAN

The purpose of this note is to make a remark that I believe sheds light on the proof of a theorem due to J. Levine [1] concerning the Arf invariant of a knot $K \subset S^3$. Before stating the theorem it is necessary to recall the definition of the Arf invariant.

Let $K \subset S^3$ be a knot and let $F \subset S^3$ be an oriented spanning surface for $K$. Then we have the Seifert pairing

$$\theta : H_1(F) \times H_1(F) \to \mathbb{Z}$$

defined by the formula $\theta(a,b) = \Lambda(a^+,b)$ where $a^+$ is the result of translating $a$ into $S^3 - F$ along the positive normal direction to $F$. $\Lambda$ denotes linking number in $S^3$. Define a symmetric bilinear pairing $<,> : H_1(F) \times H_1(F) \to \mathbb{Z}$ by the formula $<a,b> = \theta(a,b) + \theta(b,a)$, and define a mod 2 quadratic form $q : H_1(F ; \mathbb{Z}_2) \to \mathbb{Z}_2$ by the formula $q(x) \equiv \frac{1}{2}<x,x> \pmod 2$. The *Arf invariant of the knot* $K$ is the Arf invariant of this quadratic form. That is, the Arf invariant $A(K) = 0$ if $q$ takes the majority of elements of $H_1(F ; \mathbb{Z}_2)$ to $0$, and $A(K) = 1$ if $q$ takes the majority to $1$.

Let $M$ denote a matrix of $<,>$ with respect to some basis for $H_1(F ; \mathbb{Z})$. Let $D(K) = \mathrm{Det}(M)$ be the determinant of $M$. This *determinant of* $K$ is, *up to sign*, an invariant of the knot. In fact, $D(K) = \Delta(-1)$ where $\Delta(T)$ is the Alexander polynomial of the knot.

Finally, call two surfaces $F_1, F_2 \subset S^3$ *pass-equivalent* if one can be obtained from the other by ambient isotopies combined with band-passes. A *band-pass* replaces a band crossing by its mirror image crossing as illustrated in Fig. 1.

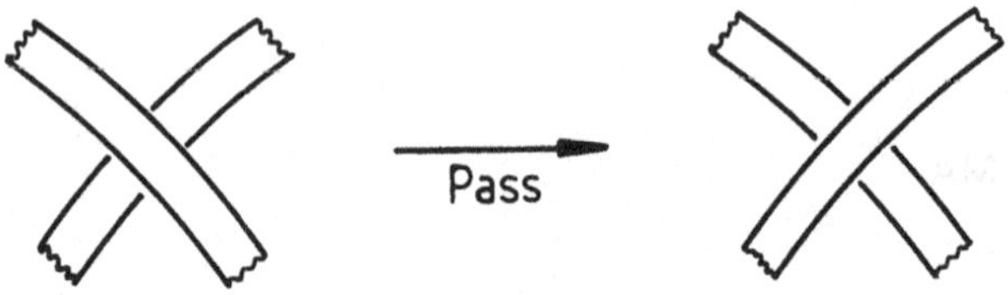


Fig. 1

LEMMA 1. *Let* $F_1, F_2 \subset S^3$ *be oriented surfaces whose respective boundaries are the knots* $K_1$ *and* $K_2$. *Let* $q_1$ *and* $q_2$ *be the mod 2 quadratic forms for* $K_1$ *and* $K_2$. *Suppose that* $F_1$ *and* $F_2$ *are pass-equivalent. Then*

*(i)* *The quadratic forms* $q_1$ *and* $q_2$ *are isomorphic. Hence* $A(K_1) = A(K_2)$.

*(ii)* $D(K_1) \equiv D(K_2) \pmod 8$.

*Proof.* Part (i) is easy. To see part (ii) note that a band-pass can change the matrix of <,> by changing a diagonal element by $\pm 4$, or by changing two entries $M_{ij}$ and $M_{ji}$ $(i \neq j)$ by $\pm 2$. Direct examination of the determinant now shows that it does not change modulo 8. For further details compare with [1] page 545.

REMARK. The proof of part (ii) of Lemma 1 is essentially the argument in Levine's paper [1]. Our point is that this algebra has a neat geometric interpretation in terms of passing bands. Band passing leaves $D(K)$ unchanged modulo 8.

LEMMA 2. *Let* $K \subset S^3$ *be a knot with oriented spanning surface* $F$. *Then* $F$ *is pass equivalent to* $F'$ *with the boundary of* $F'$ *a connected sum of trefoil knots.*

*Proof.* See [3].

THEOREM 3. (Levine). *Let* $K \subset S^3$ *be a knot. Then*

$$A(K) = 0 \Longleftrightarrow D(K) \equiv \pm 1 \pmod 8$$

$$A(K) = 1 \Longleftrightarrow D(K) \equiv \pm 3 \pmod 8 .$$

*Proof.* The matrix $M$ for the trefoil is given by $\begin{bmatrix} 2 & 1 \\ 1 & 2 \end{bmatrix}$. Hence if $T$ denotes the trefoil, then $D(T) = 3$. Lemmas 1 and

2 reduce the theorem to the case of a sum (#) of trefoils. But the Arf invariant of a connected sum of knots adds (modulo 2) while $D(K_1 \# K_2) = D(K_1)D(K_2)$. Hence the theorem follows from the fundamental formula: $3^2 \equiv 1 \pmod 8$.

REMARK. A possibly more perspicuous definition of the Arf invariant is obtained by replacing 0 by +1. That is, let $\bar{A} = +1$ when $A = 0$ and let $\bar{A} = -1$ when $A = 1$. Then Levine's Theorem becomes the formula

$$\bar{A}(K) = \left(\frac{2}{D(K)}\right)$$

where $\left(\frac{}{}\right)$ is the Legendre symbol:

$$\left(\frac{2}{D}\right) = \begin{cases} 1 & \text{if } 2 \text{ is a square } \pmod D \\ -1 & \text{if } 2 \text{ is not a square } \pmod D . \end{cases}$$ (See [2].)

In Knot theory the existence of the square root of two is decided by a majority vote.

REMARK. The ambiguity of signs $(\pm 1, \pm 3)$ in Theorem 3 is the result of the fact that $D(K)$ is only defined up to sign. Letting $V$ be a matrix for the Seifert pairing, define a *normalized determinant* $\bar{D}(K)$ by the formula

$$\bar{D}(K) = \mathrm{Det}\,((1 + \sqrt{2})V + (1 - \sqrt{2})V')$$

where $V'$ denotes the transpose of $V$. Then it can be shown that $\bar{D}(K)$ is an invariant of the knot $K$ and that it is always integral. Furthermore:

$$A(K) = 0 \iff \bar{D}(K) \equiv 1 \pmod 8$$

$$A(K) = 1 \iff \bar{D}(K) \equiv 5 \pmod 8 .$$

The author would like to thank Mark Kidwell for helpful conversations.

## REFERENCES

1. J. LEVINE, 'Polynomial invariants of knots of codimension two', *Ann. of Math.* 84 (1966) 534-554.

2. A. LIBGOBER, 'Levine's formula in knot theory and the quadratic reciprocity law'. (to appear)

3. R. ROBERTELLO, 'An invariant of knot cobordism', *Comm. Pure and Appl. Math.* 18 (1965) 543-555.

# The factorisation of knots

C. KEARTON

## 0. Introduction

It is a well known theorem of Schubert that every classical knot factorises into finitely many irreducible knots, and that this factorisation is unique up to the order of the factors. The first section of the present paper is devoted to a proof of Schubert's result. The proof presented here owes much to a group of topologists gathered at the University of Cambridge in the summer of 1979, and I am grateful to W.B.R. Lickorish for communicating their proof to me.

In higher dimensions, it is a theorem of Sosinskii that for $n \geq 3$ every n-knot factors into finitely many irreducibles, and the second section contains an outline of the proof. For $n = 3$, it is known that the factorisation is not unique, and examples are given in section three of simple 3-knots which factor in more than one way. Finally, for $n = 4q - 1$, $q \geq 2$, uniqueness of factorisation is proved for fibred simple knots whose associated quadratic form is definite.

I know of no results on the factorisation of 2-knots.

Throughout this paper, all theorems are stated for the smooth or piecewise-linear category, and in the latter case all submanifolds are assumed to be locally flat.

## 1. The classical case

If $k$ is a classical knot, $S^1 \subset S^3$, we denote its genus by $g(k)$; that is, the genus of a minimal Seifert surface of k.

PROPOSITION 1.1 (Schubert) $g(k+\ell) = g(k) + g(\ell)$ .

*Proof.* By taking the boundary connected sum of a Seifert surface $V_1$ of $k$ and $V_2$ of $\ell$, it is clear that $g(k+\ell) \leq g(k) + g(\ell)$.

Conversely, let $V$ be a Seifert surface of $k + \ell$, and let $\Sigma$ be the 2-sphere used in forming the sum $k + \ell$. Make $V$ transverse to $\Sigma$, so that $V \cap \Sigma$ consists of a single arc and some circles, the latter contained in int $V$. These circles cannot link $k + \ell$, as they lie on $V$, and so each circle bounds a disc on $\Sigma - (k+\ell)$. Starting with innermost circles, we can use the discs to perform surgeries on $V$. Each surgery disconnects $V$ or lowers its genus, and so we arrive at a Seifert surface of $k + \ell$ which has the form $V_1 \#_\partial V_2$, where $V_1$ is a Seifert surface for $k$, $V_2$ for $\ell$. Whence $g(k) + g(\ell) \le g(k+\ell)$. □

COROLLARY 1.2 *The trivial knot cannot be the sum of two non-trivial knots.*

*Proof:* $k$ is trivial if and only if $g(k) = 0$. □

LEMMA 1.3 *Let* $k = (B^3, B^1)$ *be irreducible, and let* $A$ *be an annulus properly embedded in* $B^3$*, disjoint from* $B^1$*, such that each component of* $\partial A$ *links* $k$*. Let* $X, Y$ *be the closures of the two components of* $B^3 - A$*,* $Y$ *containing* $B^1$*. Then either* $X$ *is a solid torus, or* $Y$ *is a regular neighbourhood of* $B^1$.

*Proof.* $\partial A$ bounds disjoint discs on $\partial B^3$, each one containing a point of $\partial B^1$. Let $D$ be one such disc; by regular neighbourhood theory, we can embed $D \times I$ in $B^3$ so that $D = D\times 0 = (D\times I) \cap \partial B^3$, $\partial D\times I = (D\times I) \cap X = (D\times I) \cap A$ and $(D\times I, (D\times I) \cap B^1)$ is an unknotted ball pair. Then $(A - (\partial D\times I)) \cup (D\times 1)$ is a disc properly embedded in $B^3$, and dividing $k$ into two ball-pairs. Since $k$ is irreducible, one of these must be unknotted, whence the result. □

COROLLARY 1.4 *Let* $k = (B^3, B^1)$ *be irreducible, and for* $i = 1, 2$, *let* $A_i, X_i, Y_i$ *be as in the statement of the lemma. Assume that* $X_1 \cap X_2 = \emptyset$. *Then at least one* $X_i$ *is a solid torus.*

*Proof:* If $X_1$ is not a solid torus, then $Y_1$ is a regular neighbourhood of $B^1$, and so by the construction above $X_2$ must be a solid torus, as an unknotted ball-pair cannot be of the sum of two knotted ones. □

PROPOSITION 1.5 *Let* $k$ *be an irreducible knot, and assume* $k + \ell = m + n$. *Then* $m = k + k'$ *or* $n = k + k'$ *for some knot* $k'$.

*Proof:* Let $\Sigma$ be the 2-sphere used in constructing $k + \ell$, $S$ that used in constructing $m + n$. By transversality, we can arrange that $S \cap \Sigma$ consists of a finite number of circles disjoint from the knot. From the circles which do not link the knot, choose one which is innermost on $\Sigma$. Then the discs on $\Sigma - (k+\ell)$ and $S - (k+\ell)$ which this circle bounds are disjoint except for their common boundary, and their union is a 2-sphere which by the Schoenflies theorem bounds a 3-ball in $S^3 - (k+\ell)$. So we can isotop $S$ to reduce the number of circles of $S \cap \Sigma$ which do not link the knot.

Thus we may assume that $S \cap \Sigma$ contains only circles which link the knot. Let $(B^3, B^1)$ be the ball-pair representing $k$, so that $\partial B^3 = \Sigma$. If $S \cap B^3$ contains a disc as one of its components, then this disc separates $k$ into two ball-pairs, one of which must be unknotted as $k$ is irreducible. By an isotopy of $S$, we can then reduce $n(S \cap \Sigma)$, the number of components of $S \cap \Sigma$.

So we may assume that each component of $S \cap B^3$ is an annulus. Let $A$ be such an annulus, and define $X$, $Y$ as in Lemma 1.3. Choose an $A$ such that $S \cap X = A$; thus $A$ is an "outermost" annulus as regards $B^3$. If $X$ is a solid torus, then by an ambient isotopy we can remove $A$ from the components of $S \cap B^3$, and so reduce $n(S \cap \Sigma)$. So suppose that $X$ is not a solid torus; by Lemma 1.3, $Y$ is a regular neighbourhood of $B^1$ in $B^3$. Let $A_1$ be another such "outermost" annulus; then $X_1 \subset Y$, so $X \cap X_1 = \emptyset$, and by Corollary 1.4, $X_1$ is a solid torus, and $A_1$ can be eliminated by an isotopy of $S$. Thus we can arrange for all other annuli $A_1 \subset S \cap B^3$, we have $X \subset X_1$, and $Y_1$ is a regular neighbourhood of $B^1$ in $B^3$.

Consider the two components of $S \cap \Sigma$ which are innermost on $S$: if either of these is innermost on $\Sigma$ also, then the union of the disc which it bounds on $\Sigma$ is a 2-sphere, which by the Schoenflies theorem bounds two 3-balls in $S^3$. Let $D^3$ be the 3-ball which meets $B^3$ in a disc contained in $\Sigma$. Then $D^3 \cap (\Sigma \cup S) = \partial D^3$, and $(D^3, D^3 \cap (k+\ell))$ is a proper ball-pair, representing a knot $p$ say. By pulling $p$ through $k$ (recall the proof in [2] that knot composition is commutative), we can arrange that $(D^3, D^3 \cap (k+\ell))$ is a trivial ball-pair, and hence by an isotopy we can reduce $n(S \cap \Sigma)$.

Otherwise, consider the two components $L_1$, $L_2$ of $S \cap \Sigma$ which are innermost on $\Sigma$: each of these forms part of the boundary of an annulus on $S$ contained in $\overline{S^3-B^3}$ ; these annuli $A_1$, $A_2$ must be distinct, since $L_1 \cup L_2$ is the boundary of an annulus on $S$ contained in $B^3$. Let $\Delta_1, \Delta_2$ be the two disjoint discs on $\Sigma$ such that $X \cap (\Delta_1 \cup \Delta_2) = \partial\Delta_1 \cup \partial\Delta_2$, and $L_i \subset \Delta_i$. Then for at least one of the $A_i$ we have $\partial A_i \subset \Delta_i$; let us assume that $\partial A_1 \subset \Delta_1$. Then $\partial A_1$ bounds an annulus $B_1 \subset \Delta_1$, and $A_1 \cup B_1$ is a torus. Let $Z_1$ be the component of $S^3$ bounded by $A_1 \cup B_1$, with $Z_1 \cap B^3 = B_1$. Then $Z_1 \cap S$ consists of a number of annuli (including $A_1$). Pick the "innermost" annulus $A'$, by which we mean the annulus $A'$ such that $\partial A' = \partial B'$ where $B'$ is an annulus in $B_1$, $\partial Z' = A' \cup B'$, $Z' \cap B^3 = B'$, $Z' \cap S = A'$, $Z'$ is nearest to k. See Fig. 1.

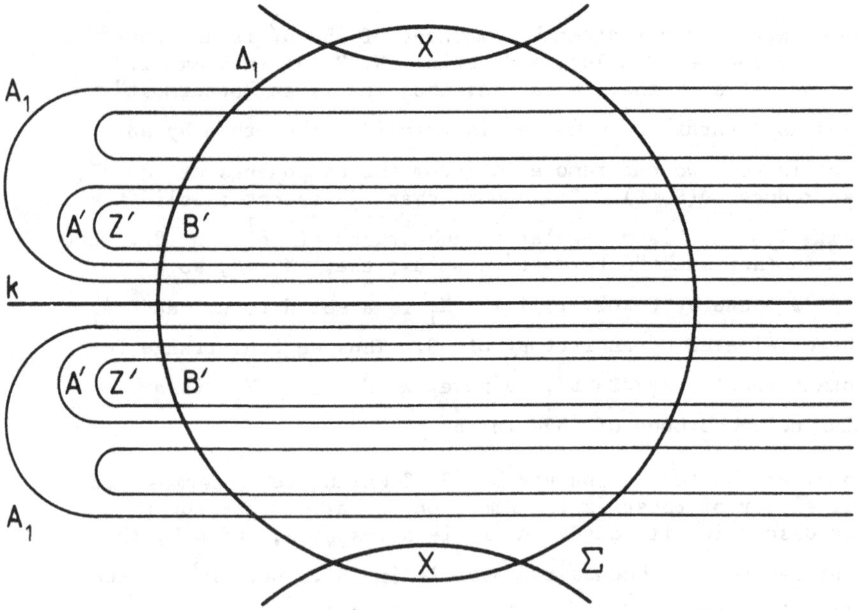


Fig.1

The inner component of $\partial A'$ bounds a disc $\Delta' \subset \Delta_1$; let $B'^3$ be a regular neighbourhood of $\Delta' \cup Z'$ in $\overline{S^3-B^3}$. Then $(B'^3, B'^3 \cap (k+\ell))$ is a proper ball-pair, representing a knot p say. $B'^3 \cap A_1$ is the boundary of a regular neighbourhood of

$B'^3 \cap (k+\ell))$ and similarly for the other annuli meeting $B'^3$ (except $A'$). Pulling $p$ through $k$, as above, leaves the picture as before, but with $Z'$ replaced by a solid torus. By an isotopy we can then reduce $n(S \cap \Sigma)$.

Thus in all cases we may by an ambient isotopy reduce $n(S \cap \Sigma)$, and so we may assume that $S \cap \Sigma$ is empty.

If $S \cap B^3$ is empty, then $(B^3, k)$ is contained in one of the ball-pairs representing $m$ or $n$; let us assume it is the ball-pair $(D^3, m)$. Choose an unknotted ball-pair $(C^3, t)$ such that $C^3 \cap D^3$ is a disc in $\partial C^3$ and $\partial D^3$. As in Fox's proof of the commutativity of knot composition, we can slide $(B^3, k)$ along the knot until $(B^3, k)$ replaces $(C^3, t)$. Clearly then $m = k+k'$, where $k'$ is represented by the ball-pair $(\overline{D^3-B^3}, \overline{(D^3-B^3)} \cap m)$.

If $S \cap B^3$ is not empty, then $S \subset B^3$, and we can assume that the ball-pair $(D^3, m)$ representing $m$ is contained in $(B^3, k)$. A similar argument to the one above shows that $k = m + m'$; since $k$ is irreducible, either $k = m$ and $m'$ is trivial, or $k = m'$, $m$ is trivial, and $n = k + \ell$. □

THEOREM 1.6 *Every classical knot factors into finitely many irreducible knots, and this factorisation is unique up to the order of the factors.*

*Proof:* Finite factorisation follows at once from Proposition 1.1, and the fact that $k$ is trivial if and only if $g(k) = 0$.

Unique factorisation follows from Proposition 1.5, which states that "irreducible implies prime", in the usual meaning of those words. □

## 2. Finite Factorisation in Higher Dimensions

We shall outline Sosinskii's [7] proof of finite factorisation into irreducibles of n-knots, for $n \geq 3$. An n-knot is a pair $(S^{n+2}, S^n)$; by excising the interior of a regular (or tubular) neighbourhood of a point on $S^n$, an n-knot can be represented by a knotted ball-pair $(B^{n+2}, B^n)$, where the boundary is a trivial $(n-1)$-knot. If $k$ and $\ell$ are two n-knots represented by two such ball-pairs, then gluing the ball-pairs together along the boundary so that the orientations match up gives an n-knot $k + \ell$.

Let $k$ be an n-knot, and $K = \overline{S^{n+2}-S^n \times B^2}$ the closed complement of a regular neighbourhood of $k$. $K$ is the *exterior* of k. Let V be a Seifert surface of k; that is, an orientable compact (n+1)-manifold contained in $S^{n+2}$ with $\partial V = S^n$. Let X denote K split open along V. Then $\pi_1(X)$ is finitely-presented, say by $\langle x_1,\ldots,x_n : r_1,\ldots,r_m\rangle$, and $\pi_1(K)$ is finitely-presented by $\langle x, x_1,\ldots,x_n : r_1,\ldots,r_m, s_1,\ldots,s_p\rangle$ for some $s_1,\ldots,s_p$. Let $\pi_1(k)$ denote $\pi_1(K)$. If $h: \pi_1(K) \to H_1(K) = \langle t:\rangle$ is the Hurewicz map, then $h(x) = t$ and $h(x_i) = 1$ for each i. We call $x, x_1,\ldots,x_n$ a *canonical* set of generators for $\pi_1(k)$ when they have the latter property and n is a minimum.

PROPOSITION 2.1 (Sosinskii) *Let* k, $\ell$ *be* n*-knots and let* $x, x_1,\ldots,x_m$ *be a canonical set of generators for* $\pi_1(k)$, $y, y_1,\ldots,y_n$ *a canonical set of generators for* $\pi_1(\ell)$. *Then* $x, x_1,\ldots,x_m, y_1,\ldots,y_n$ *is a canonical set of generators for* $\pi_1(k+\ell) \cong \pi_1(k) \underset{\langle t:\rangle}{*} \pi_1(\ell)$.

The proof is omitted.

THEOREM 2.2 (Sosinskii) *Let* k *be an* n*-knot*, $n \geq 3$. *Then* k *can be written as a sum of finitely many irreducible knots.*

*Proof:* By Proposition 2.1, k can be written as $k_1 + \ldots + k_r + \ell$ where $\pi_1(\ell) \cong \langle t:\rangle$ and each $k_i$ has the property that if $k_i = m_i + n_i$, then $\pi_1(m_i)$ or $\pi_1(n_i) \cong \langle t:\rangle$. Now consider the exterior K of k, and let $\tilde{K}$ be the infinite cyclic cover of K. $H_2(\tilde{K})$ is a finitely-presented module over $\Lambda = \mathbb{Z}[t,t^{-1}]$, and $H_2(\tilde{K}:Q)$ is a finitely-presented module over $\Gamma = Q[t,t^{-1}]$. In each case the modules are $\Lambda$- or $\Gamma$- torsion-modules. Since $\Gamma$ is a PID, $H_2(\tilde{K}:Q)$ can be written uniquely as a sum of cyclic p-primary modules, where p denotes a prime in $\Gamma$. Let $r_2(k)$ denote the number of such summands. If $k = m+n$, then $H_2(\tilde{K}:Q) \cong H_2(\tilde{M}:Q) \oplus H_2(\tilde{N}:Q)$, and so $r_2(k) = r_2(m) + r_2(n)$. Thus k can be written as a sum $k_1 + .. + k_r + \ell$ (to save on notation, these are not the same $k_i$, $\ell$ as above) such that $\pi_1(\ell) \cong \langle t:\rangle$, $H_2(\tilde{L}:Q) = 0$, and each $k_i$ has the property that if $k_i = m_i + n_i$, then (say) $\pi_1(n_i) \cong \langle t:\rangle$ and $H_2(\tilde{N}_i:Q) = 0$.

Let $T_2(\tilde{K})$ denote the $\mathbb{Z}$-torsion-submodule of $H_2(\tilde{K})$: by Lemme II.8 of [4], $T_2(\tilde{K})$ is a finite group. Since $\mathbb{Z}$ is a PID, we can apply the argument above to the $\mathbb{Z}$-module $T_2(\tilde{K})$ to obtain $k = k_1 + .. + k_r + \ell$ such that $\pi_1(\ell) = \langle t:\rangle$, $H_2(\tilde{L}) = 0$, and each $k_i$ has the property that if $k_i = m_i + n_i$, then (say) $\pi_1(n_i) = \langle t:\rangle$ and $H_2(\tilde{N}_i) = 0$.

This process may be repeated on $H_q(\tilde{K})$ until $q = [(n+1)/2]$, to obtain $k = k_1 + .. + k_r + \ell$ such that $\pi_1(\ell) = \langle t:\rangle$, $H_q(\tilde{L}) = 0$ for $1 \leq q \leq [(n+1)/2]$, and each $k_i$ has the property that if $k_i = m_i + n_i$ then $\pi_1(n_i) = \langle t:\rangle$ and $H_q(\tilde{N}_i) = 0$ for $1 \leq q \leq [(n+1)/2]$. By duality, Hurewicz, and covering space theorems, it follows that L and $N_i$ have the homotopy type of $S^1$. By theorems of Levine [5], $\ell$ and $n_i$ are unknotted, and so the $k_i$ are irreducible. □

## 3. Non-unique Factorisation for 3-knots

The results of this section appear in [3].

For any integer $\alpha$, define

$$A_\alpha = \begin{pmatrix} 1 & 0 & -1 & 0 & 0 & 0 & 0 & 0 \\ 0 & 1 & \alpha & -1 & 0 & 0 & 0 & 0 \\ 0 & -\alpha & 1 & -1 & 0 & 0 & 0 & 0 \\ 0 & 0 & 0 & 1 & -1 & 0 & 0 & 0 \\ 0 & 0 & 0 & 0 & 1 & -1 & 0 & 0 \\ 0 & 0 & 0 & 0 & 0 & 1 & -1 & 0 \\ 0 & 0 & 0 & 0 & 0 & 0 & 1 & -1 \\ 0 & 0 & 0 & 0 & 0 & 0 & 0 & 1 \end{pmatrix}$$

Denoting the transpose by $A'_\alpha$, the matrix $A_\alpha + A'_\alpha$ is the same as that which appears in [6; p 51]. It is easy to check that

(i) $\det(A_\alpha + A'_\alpha) = 1$

(ii) $\text{signature}\,(A_\alpha + A'_\alpha) = 8$

(iii) $\det A_\alpha = 1 + \alpha^2$

Define $A_{\alpha,\beta} = \begin{pmatrix} A_\alpha & 0 \\ 0 & A_\beta \end{pmatrix}$, and note that

(iv) $\det(A_{\alpha,\beta} + A'_{\alpha,\beta}) = 1$

(v) signature $(A_{\alpha,\beta} + A'_{\alpha,\beta}) = 16$

(vi) det $A_{\alpha,\beta} = (1+\alpha^2)(1+\beta^2)$ .

By the results of Levina [5], $A_{\alpha,\beta}$ is S-equivalent to a Seifert matrix of a simple 3-knot $k_{\alpha,\beta}$, and $k_{\alpha,\beta}$ is uniquely determined by $A_{\alpha,\beta}$; this uses properties (iv) and (v). By (vi), the leading coefficient of the Alexander polynomial of $k_{\alpha,\beta}$ is $(1+\alpha^2)(1+\beta^2)$. Since every 3-knot must have signature a multiple of 16, and $A_{\alpha,\beta}$ is a 16 × 16 matrix, it follows from [5] that $k_{\alpha,\beta}$ is irreducible.

Take $\alpha,\beta,\gamma,\delta$ to be distinct positive integers. By [5; theorem 3] , $k_{\alpha,\beta} + k_{\gamma,\delta} = k_{\alpha,\gamma} + k_{\beta,\delta}$ , the Seifert matrices being S-equivalent. The knot $k_{\alpha,\beta}$ is irreducible and distinct from the irreducible knots $k_{\alpha,\gamma}$ and $k_{\beta,\delta}$ since the Alexander polynomials differ in the leading coefficient. We have therefore constructed a 3-knot which factors into irreducibles in more than one way.

## 4. Some Unique Factorisation

Consider a simple (4q-1)-knot k; that is, K has the homotopy (2q-1)-type of a circle. Assume $q \geq 2$, and let A be a non-singular Seifert matrix of k. It is a result of Trotter [8] that a non-singular Seifert matrix exists, and that any two such Seifert matrices of a knot k are congruent by unimodular matrices over the ring $\mathbb{Z}[1/d]$ where d = det A. We shall restrict our attention to the case det A = 1, that is to fibred knots.

Then the matrix A + A' yields a quadratic form on $H_{2q}(\tilde{K})$ , and $H_{2q}(\tilde{K}) = \oplus_1^{2g} \mathbb{Z}$ where A is a 2g × 2g matrix. Moreover, the map represented by $A^{-1}A'$ is an isometry of this quadratic lattice. If the quadratic form is (positive or negative) definite, then we call the knot k *definite*.

THEOREM 4.1 *Let* k *be a simple* (4q-1)-*knot,* $q \geq 2$ *which is fibred and definite. Then* k *factors uniquely into irreducibles.*

*Proof:* The quadratic lattice $H_{2q}(\tilde{K})$ can be written as an orthogonal direct sum of indecomposable sublattices, and this decomposition is unique (see [1: p 363]). Say $H_{2q}(\tilde{K}) = L_1 \perp \ldots \perp L_n$. The isometry represented by $A^{-1}A'$ must therefore permute the $L_i$, and so $H_{2q}(\tilde{K})$ splits uniquely as an orthogonal direct sum $H_1 \perp \ldots \perp H_m$ of irreducible $\mathbb{Z}[t,t^{-1}]$-modules, where $t$ is the map represented by $A^{-1}A'$.

Setting $S = A + A'$ and $T = A^{-1}A'$, we see that $A = S(I + T)^{-1}$. Thus choosing a $\mathbb{Z}$-basis for each $H_i$, we see that $A$ is congruent by a unimodular matrix over $\mathbb{Z}$ to one of the form $A_1 \oplus \ldots \oplus A_m$. Moreover, each $A_i$ satisfies $\det(A_i + A_i') = \pm 1$, and so $A_i$ is a Seifert matrix of a unique simple $(4q-1)$-knot $k_i$ by [5]. Since $H_i$ is irreducible, so is $k_i$, and since $H_i$ is unique, so too is $k_i$. □

## 6. Stop Press

Since this article was submitted, Eva Bayer has found examples of non-unique factorisation in the following dimensions.

(i) There exist simple $(2q-1)$-knots, for every $q > 2$, which factorise into irreducibles in more than one way.

(ii) There exist simple $2q$-knots, for every $q \geq 4$, which factorise into irreducibles in more than one way.

In each case specific examples are given, and details may be found in her preprint "Factorisation is not unique for higher dimensional knots".

By contrast, J.A. Hillman has shown that certain $2q$-knots have a unique factorisation into irreducibles. His results are contained in the preprint "Finite knot modules and the factorisation of certain simple knots".

## References

1. J.W.S. CASSELS, *Rational Quadratic Forms*, Academic Press, L.M.S. Monograph No. 13 (1978).

2. R.H. FOX, *A quick trip through knot theory. Topology of 3-manifolds*, Ed. M.K. Fort. Prentice-Hall Inc. (1962).

3. C. KEARTON. Factorisation is not unique for 3-knots. *Indiana Univ. Math. Jour.* 28 (1979), 451-452.

4. M.A. KERVAIRE, Les Noeuds de dimensions superieures. *Bull. Soc. Math. France*, 93 (1965), 225-271.

5. J. LEVINE, An algebraic classification of some knots of codimension two. *Comm. Math. Helv.* 45 (1970), 185-198.

6. J.P. SERRE, *A Course in Arithmetic*. Springer-Verlag (1973).

7. A.B. SOSINSKII, Decompositions of knots. *Math. USSR Sbornik*, 10 (1970), 139-150.

8. H.F. TROTTER, Homology of group systems with applications to knot theory. *Ann. of Math.* 76 (1962), 464-498.

*Added in proof*. An example of T. Maeda [On a composition of knot groups II - algebraic bridge index; Math. Sem. Notes Kobe Univ. 5 (1977), 457-464] indicates that Proposition 2.1 is false. The proof of Theorem 2.2 is still valid when $\pi_1(K)$ is infinite cyclic, but the general result appears to be an open question.

# Seven excellent knots

R. RILEY

The excellent hyperbolic structure that W. Thurston has now shown to exist on all suitable knot complements is still rather difficult to describe or write down explicitly in specific cases. There are published examples of this structure (exactly one knot and a large but restricted collection of links), but these all relied on the most exceptional feature that the rings $R(\theta)$ of the excellent p-reps concerned were always *discrete* rings of algebraic integers. In a companion paper [16] we shall discuss methods of proving that an explicitly given discrete subgroup $G$ of $PSL(2, \mathbb{C})$ really is discrete when the definition of $G$ did not make its discreteness clear. This situation often arises when the group is defined algebraically in relation to the solution of certain polynomial equations, and the matrix entries of $G$ generate an algebraic number field of high degree. The methods of [16] were developed specifically for the application to groups $G$ of the type considered in this paper, and the present paper began life as a segment of [16], which thereby grew so large that it had to fission. We shall exhibit the excellent hyperbolic structure most explicitly for seven typical knots which exhibit the greatest range of behaviour that our small supply of worked examples permits. We shall not rely on a rare and exceptional feature such as the discreteness of $R(\theta)$, and the results below would have provided a powerful stimulus for the discovery of a general existence theorem for hyperbolic structure on suitable 3-manifolds had Professor Thurston not been so quick.

This paper is organized as follows. In section 1 we restate our basic setup for knots and hyperbolic geometry, and especially we include an account of the longitude polynomial taken from an unpublished paper. In section 2 we repeat or expand on our formalism for 2-bridge knots and prepare for section 3. This section is a detailed account of the excellent hyperbolic structure for the three 2-bridge knots $5_2 = (7, 3)$, $7_7 = (21, 13)$, and $6_3 = (13, 5)$. In section 4 we summarize our piles of computer printout and rolls of diagrams of Ford domains for several dozen further examples. These strongly suggest that there are just 3 patterns for all the excellent 2-bridge knots and that our single example for each pattern is really typical. We also discuss conjectures for these knots that no one would have dared make on the basis of just 3 examples.

In section 5 we give an account of the excellent hyperbolic structure for the four 3-bridge knots $8_{21}$, $9_{43}$, $9_{35}$, and $9_{48}$. We need more examples because the 3-bridge knots are much less predictable than 2-bridge knots. We have omitted an account of the non-invertible knot $8_{17}$ where the calculations have the same completeness as for $9_{43}$, because the field degree $[F(\theta):\mathbb{Q}] = 18$ for the excellent p-rep $\theta$ and the diagram of the Ford domain would probably not be copied well by the duplicating process. Our last examples, $9_{35}$ and $9_{48}$, are especially interesting because their p-rep sets are algebraically very similar and we do not understand why. Their homeotopy groups are both isomorphic to the dihedral group $D_6$, and this is no surprise for $9_{35}$ because all the symmetries are easily seen in a diagram of the knot. However, $9_{48}$ has a peculiar symmetry of period 3 which has no fixed points in $S^3$ and is definitely not a rotation. We shall discuss this symmetry in detail because it is not easily seen and because there doesn't seem to be any other published or privately distributed account of a similar knot symmetry. M. Boileau and L. Siebenmann have developed an entirely different method for calculating the homeotopy groups of "algebraic knots" in the sense of Conway [2], and used it for the 3-bridge knots with crossing number $\leq 9$. Our results for $9_{48}$ agree, but this author once had the strange idea that $9_{48}$ was the pretzel knot $(3, 3, -3)$ (which actually is $9_{42}$). This generated a short lived controversy about $9_{48}$ that is now happily settled.

The work summarized by this paper began about 1970 and continued intermitently in various phases right up to early 1980. The first two years work was done during temporary lectureships at Southampton University. In the first half of 1973 I visited the Université de Strasbourg with the financial support of the C.N.R.S. and used their computer facilities to complete the purely algebraic phase of the 3-bridge knot calculations. During the subsequent 2½ years of unemployment, I worked out the methods of [16] and used them to discover the first examples of excellent hyperbolic structure on knot and link complements in $S^3$. The present paper is actually a replacement for the unpublished paper CPG [13] written during this period where the excellence of $5_2$ and $7_4$ was proved. We shall demonstrate below what can be done with the methods of CPG when these are implemented in computer programs. This implementation and all subsequent analysis was carried out while the author was a research fellow at Southampton University during the four years 1976 - 1980 with the financial support of the Science Research Council. I am deeply grateful to the project supervisor, Dr. David Singerman, for his essential assistance in arranging this financial support and to Southampton University for the excellent facilities provided. This paper was finally completed at the Institute for Advanced Study in Princeton with the help of financial support from the N.S.F.

During the three months following the completion of the first version of the present paper there were several further developments. First, F. Bonahon and L. Siebenmann gave us preprints of articles that are good background for this paper. In particular, we are very happy to refer to J. Montesinos [8] for a discussion of Conway's rational tangles and 2-bridge knots supporting §4 below. Second, Bonahon supplied Figure 7(e) illustrating the strange symmetry of $9_{48}$ and the news that all our results for 3-bridge knots in §5 below agree with secret calculations of M. Boileau and L. Siebenmann. Then, in addition to the official referee, F. Bonahon and J. Lannes read our preprint critically and offered many suggestions for improvements. Some of these were taken up and others were impractical, but the main complaint of all referees and editors was that this paper relies too completely on a close reading of our Elliptical Path [15]. However, [15] is our general survey article on this subject and needs only minor corrections to become the best survey we could do today, whereas the present paper is a more specialized continuation of [15]. In particular, §1 below develops material briefly sketched in §4 of [15], and one should certainly study the easy example of the Borromean rings in §3 of [15] before worrying about the hard examples in §§3, 5 below. Furthermore, if [16] survives its hazardous journey to publication, it will contain a restatement of our general setup for hyperbolic geometry and Kleinian groups, in addition to an account of the methods applied in this paper. Finally, in May 1980 we at last got around to using the better than 33 decimal figure floating point arithmetic available on the Princeton University computer to push the algebraic part of the calculations of p-reps for the knots $9_{43}$, $9_{44}$ and $8_{17}$ one step further. The results appear to be successful for $9_{43}$ and $9_{44}$ (see §5 for $9_{43}$), but our attempt failed again for $8_{17}$. It seems that we shall really need more than 35 decimal figure accuracy for $8_{17}$, and we dread to think what will be required to work difficult examples like $9_{32}$, $9_{33}$.

## 1. The basic setup

We begin by recalling and elaborating our account of the automorphism group of an excellent knot group in relation to the excellent hyperbolic structure on the knot complement from [15]. Our standard notation for knots, p-reps, Pell groups, etc. that we explained in [12 , 14 , 15] will mostly be taken for granted.

We shall frequently refer to the standard parabolic elements $A\{t\}$ , $A$ and $B\{t\}$ in $PSL(\mathbb{C})$ , which are

$$A\{t\} = \begin{bmatrix} 1 & t \\ 0 & 1 \end{bmatrix} , \; A = A\{1\} , \; B\{t\} = \begin{bmatrix} 1 & 0 \\ -t & 1 \end{bmatrix} .$$

Also we abbreviate $B\{t\}$ to $B$ as soon as the context makes the parameter unambiguous. In general, we denote parameters $t , u , \ldots$ by $\{t , u , \ldots\}$ and arguments by $(\;)$ , thus $A(t) = t + 1$ . Another hyperbolic isometry that we shall need is the EH-rotation

$$R\{t\} = R_t = \begin{bmatrix} i & -it \\ 0 & -i \end{bmatrix} : z \longrightarrow -z+t \quad \text{and} \quad (z , h) \in U_3 \longrightarrow (R_t(z) , h).$$

This too is quickly abbreviated to $R$ . Note that

$$R_t^2 = E , \; R_t A\{u\} R_t^{-1} = A\{u\}^{-1} , \; R_t B\{u\} R_t^{-1} = \begin{bmatrix} 1+tu & -t^2u \\ u & 1-tu \end{bmatrix} .$$

In particular, $R_0 B\{u\} R_0 = B\{u\}^{-1}$ , so $R_0$ normalizes every group $\langle A , B\{u\} \rangle$ .

Let $G$ be a Pell group (meaning that the orbit space $U_3/G$ has finite volume) and let $\psi$ be an automorphism of $G$ . Then A. Marden's version [7] of Mostow's Rigidity Theorem (hereafter MRT) asserts that there exists $T \in P\Gamma L(\mathbb{C})$ such that

$$\psi(x) = T \, x \, T^{-1}$$

for all $x \in G$ . Only the identity $E$ commutes with every element of a Pell group, so this $T$ is unique. Therefore the automorphism group $Aut(G)$ is naturally identified with a subgroup $\mathcal{A} \subset P\Gamma L(\mathbb{C})$ , and $G$ is naturally identified with the subgroup of inner automorphisms of $G$ . Hence $G$ is a normal subgroup of $\mathcal{A}$ and of $\mathcal{A}^+ := \mathcal{A} \cap PSL(\mathbb{C})$ , the subgroup of orientation-preserving transformations of $\mathcal{A}$ . It is very easy to see that $\mathcal{A}^+$ and $\mathcal{A}$ are discrete groups, that their orbit spaces have finite volume, and that $G$ has finite index in both groups. Hence

$$\mathrm{Out}^+(G) := A^+(G) \quad \text{and} \quad \mathrm{Out}(G) := A/G$$

are finite groups. When $G$ is torsion-free the orbit space $M(G) = U_3/G$ is an irreducible 3-manifold. Also if $M(G)$ is not compact then it is sufficiently large, and a theorem of F. Waldhausen [19] implies that $\mathrm{Out}(G)$ is naturally identified with the homeotopy group of $M(G)$.

Let $k \subset S^3$ be an oriented prime knot. We denote both the type and the knot exterior (also called the knot space) by $K$, and the knot group $\pi K = \pi_1(K ; *) = \pi_1(S^3 - k ; *)$. A peripheral subgroup $P$ of $\pi K$ is generated by an oriented meridian $x$ and an oriented longitude $\gamma$, and $P = \langle x , \gamma \rangle$ is isomorphic to $\mathbb{Z} \oplus \mathbb{Z}$. If $P' = \langle x' , \gamma' \rangle$ is a second peripheral subgroup there is an inner automorphism $\eta$ such that $\eta(x') = x$, $\eta(\gamma') = \gamma$. Hence if $\beta'$ is any automorphism of $\pi K$ there is an inner automorphism $\eta$ such that $\beta := \eta\beta'$ leaves $P$ setwise fixed. The longitude $\gamma$ and its inverse are the only non-trivial elements of $P$ which are not proper powers of other elements and which belong to the commutator subgroup of $\pi K$, so necessarily $\beta(\gamma) = \gamma^\delta$ where $\delta = \pm 1$. It is easy to see that $\beta(x)$ has the shape $(x\gamma^{-e})^\varepsilon$ where $\varepsilon = \pm 1$ and $e$ is an integer. We call $e$ the *incoherence* of $\beta$, and we call $\beta$ *coherent* when $e = 0$. We also call $K$ *coherent* when all automorphisms of $\pi K$ are coherent. J. Montesinos pointed out that this definition of the incoherence implicitly refers to $k \subset S^3$ and not to $S^3 - k$ alone, for the following reason. Let $N$ be a closed 3-manifold formed by Dehn surgery along $k$, so that there is a knot $k_1$ in $N$ such that $N - k_1 \approx S^3 - k$. Let $x_1$ be a meridian for $k_1$, then $x_1 = x^a\gamma^b$ for certain integers $a , b$, and we shall assume $a > 0$. Then $\beta(x_1) = (x_1\gamma^{-f})^\varepsilon$, where we compute

$$f = (1 - \varepsilon\delta)b + ae .$$

Call $f$ the *incoherence* of $\beta$ for $N$. When $\delta = \varepsilon$ we see that $\beta$ is coherent for $S^3$ if and only if $\beta$ is coherent for $N$. When $\delta = -\varepsilon$ then $f = 2b + ae$, and if also $e = 0$ then $\beta$ is coherent for $N$ only when $b = 0$. In these circumstances $a = 1$ and $N = S^3$.

Suppose $\beta'$ is induced by an autohomeomorphism $\alpha$ of $S^3 - k$ which fixes the basepoint of $\pi K$, and consider the numbers $\varepsilon , \delta$ for $\beta'$ and $\beta$ of the preceding paragraph. Then $\varepsilon = \delta$ when $\alpha$ preserves the orientation of $S^3 - k$ and $\varepsilon = -\delta$

otherwise. For the knot exterior K has a torus T as boundary, and an orientation of K induces an orientation of T and vice versa. We can arrange that $\alpha(T) = T$, so $\alpha(K) = K$, and then $\varepsilon = \delta$ means that $\alpha$ preserves the orientation of T. This makes the assertion obvious. Also note that $\beta'$, $\beta$ are coherent if and only if $\alpha$ is the restriction of an autohomeomorphism of $S^3$. In general we call an autohomeomorphism of $N - k_1$ *coherent* or *incoherent* (for N) according as it is or is not the restriction of an autohomeomorphism of N.

Now assume further that k is excellent and let $\theta : \pi K \to \pi K\theta$ be an excellent p-rep. Because $\pi K\theta$ is a Pell group $\mathrm{Aut}(\pi K)$ is naturally isomorphic to the normalizer $A$ of $\pi K\theta$ in $P\Gamma L(\mathbb{C})$. We may suppose $x\theta = A$, where $x, \gamma$ is the selected oriented meridian, longitude pair of the above discussion. Then $(\pi K\theta)_\infty = \langle x, \gamma\rangle\theta$ is the subgroup of EH-translations of $\pi K\theta$, and $\gamma\theta = A\{g\}$ where the complex number g is called the *longitude parameter* of K. We deduce from the above discussion that $A$ is generated by $\pi K\theta$ and by the generators of $A_\infty$ which are not already in $(\pi K\theta)_\infty$. One or two of these generators can have the shape $A\{(m + ng)/r\}$ where $m, n, r \in \mathbb{Z}$, $r \geq 2$, and $\gcd(m, n, r) = 1$. Another can be an EH-rotation $R_t$ of order 2. (There cannot be an EH-rotation of any higher order because $\gamma\theta$ must either be preserved or inverted). Finally, there might be a glide-reflection $J = J\{a, b\}$ whose action is

$$J : z \to a\bar{z} + b \ (a, b \in \mathbb{C}, |a| = 1) \ , \ J : (z, h) \to (J(z), h) \ .$$

Conjugation by such transformations has the following effect on $(\pi K\theta)_\infty$. The EH-translations leave every element fixed, the EH-rotations invert every element, and the glide-reflection preserves the elements of one cyclic subgroup while inverting those of another. The autohomeomorphisms for the first two kinds of transformations preserve the orientation of $S^3 - k$ and are clearly coherent. The glide-reflection $J\{a, b\}$ corresponds to an orientation-reversing autohomeomorphism of $S^3 - k$ and is coherent if and only if $a = \pm 1$. The EH-translations preserve the knot orientation, the EH-rotations reverse it, and the coherent glide-reflections reverse or preserve it according as $a = +1$ or $a = -1$. Only the EH-rotations $R_t$ have an obvious standard geometric interpretation, viz. they correspond to rotations of $180^\circ$ about an axis (a circle in $S^3$) which intersects the knot. If k is symmetric about an axis not meeting it the corresponding rotation will correspond to some $A\{(m + ng)/r\}$, and we expect that $m = 0$ in all such cases. We shall exhibit an automorphism $A\{(m + ng)/r\}$ whose geometric interpretation is rather non-obvious, and an incoherent glide-reflection would be especially interesting because it would contradict the Property P conjecture. Note, however, that this discussion applies virtually without change to an excellent knot in any orientable closed 3-manifold, and none of the possible geometric interpretations would really be weird in that context.

The *longitude polynomial* $Q(y)$ is the primitive irreducible polynomial in $\mathbb{Z}[y]$ such that $Q(g) = 0$ and $q(0) > 0$. This polynomial is the most important computable invariant of knot type that has yet emerged from the excellent hyperbolic structure, and it might by itself decide which kinds of glide-reflections normalize $\pi K\theta$.

THEOREM 1. *In the situation under discussion suppose* $\pi K\theta$ *is normalized by a glide-reflection* $J$ *with incoherence* $e$. *Then the longitude polynomial* $Q$ *satisfies the functional equation*

$$Q(y) = (ey - 1)^r \, Q\left(\frac{y}{ey-1}\right), \quad \textit{where} \quad r = \deg Q(y) \,. \qquad (1.1)$$

*Furthermore* $r$ *is even and the longitude polynomial alone determines the incoherence* $e$. *In particular, all the orientation-reversing autohomeomorphisms of* $S^3 - k$ *have the same incoherence. Finally, if* $g$ *is an algebraic integer and* $e \neq 0$ *then*

$$e^r Q\left(\frac{1}{e}\right) = \pm 1 \,.$$

*Proof.* Suppose $J = J\{a, b\}$, then one calculates that

$$J^{-1} = J\{\overline{a^{-1}}, \overline{-a^{-1}b}\}, \quad \text{so} \quad JA\{t\}J^{-1} = A\{a\bar{t}\} \,.$$

Because $J$ corresponds to an automorphism $\beta$ of $\pi K$ such that $\beta(x) = (x\gamma^{-e})^{\varepsilon}$ we get $a = \varepsilon(1 - eg)$. From $\beta(\gamma) = \gamma^{-\varepsilon}$ we then get

$$-\varepsilon g = \varepsilon(1 - eg)\bar{g}\,, \quad \text{whence} \quad \bar{g} = \frac{g}{eg-1} \,. \qquad (1.2)$$

Let $\tilde{Q}(y) := (ey-1)^r \, Q\left(\frac{y}{ey-1}\right)$, then $\tilde{Q}(g) = (eg-1)^r \, Q(\bar{g}) = 0$.

Hence $\tilde{Q}(y)$ is a constant multiple of $Q(y)$, and from $Q(0) = (-1)^r \, \tilde{Q}(0)$ we get $Q(y) = (-1)^r \, \tilde{Q}(y)$. Let

$$I : \mathbb{P}^1(\mathbb{C}) \longrightarrow \mathbb{P}^1(\mathbb{C}) \,, \quad y \longmapsto \frac{y}{ey-1} \,,$$

then $I$ is an involution and $I$ maps $\mathbb{P}^1(\mathbb{R})$ on itself. The fixed points of I are $0, 2/e$, and these are not roots of $Q(y) = 0$ because they are rational whereas $Q$ is irreducible. We have just proved that $\rho \to I(\rho)$ is a permutation of the roots $\rho$ of $Q(y) = 0$. If the degree $r$ of $Q$ were odd then the number of real $\rho$ would also be odd, whence at least one root would satisfy $\rho = I(\rho)$. This contradiction shows that $r$ is even and that the functional equation (1.1) holds. Write

$$Q(y) = \sum c_\nu y^\nu ,$$

then

$$(ey-1)^r Q\left(\frac{y}{ey-1}\right) = c_0(ey-1)^r + c_1(ey-1)^{r-1}y + \ldots + c_r y^r$$

$$= c_0 - [erc_0 + c_1]y + \ldots + [c_r + ec_{r-1} + \ldots + e^r c_0]y^r = Q(y).$$

Hence $2c_1 + erc_0 = 0$, and because $c_0 > 0$ the value of $e$ is fixed by the coefficients $c_0, c_1$ of $Q$ alone. Finally, when $e \neq 0$ comparing the coefficients of $y^r$ gives $e^r Q(e^{-1}) = c_r$, and when $g$ is an algebraic integer $c_r = \pm 1$. This completes the proof of the theorem. □

The need for the longitude polynomial to satisfy such a functional equation is such a stiff necessary condition for $S^3 - k$ to admit an orientation-reversing autohomeomorphism that we wonder whether it is also a sufficient condition. R. Hartley and A. Kawauchi have jointly [4] and separately (preprints) shown that the Alexander polynomial of an amphicheiral knot has certain special properties. Perhaps these should be considered as analogous to (1.1). Note that when $J$ above is coherent then $g$ is pure imaginary.

The longitude polynomial also influences the kind of EH-translations that might normalize $\pi K\theta$. Suppose that $\theta$ has $x\theta = A$ and also there is a second meridian $x_2$ of $\pi K$ such that $x_2\theta = B\{\xi\}$ for some $\xi \in \mathbb{C}$. Then we say that $\theta$ is a *normalized* p-rep. A normalized p-rep $\theta$ such that $R(\theta)$ is a ring of algebraic integers is called an *integral* p-rep. Every p-rep of $\pi K$ is simply equivalent to a normalized p-rep, and we have not yet found a p-rep which is not simply equivalent to an integral p-rep.

THEOREM 2. *Suppose* $\theta$ *is an excellent normalized p-rep of* $\pi K$ *and suppose also that* $\pi K\theta$ *is normalized by* $A\{\eta\}$ *for some* $\eta \in \mathbb{C}$. *Then* $\eta \in R(\theta)$. *In particular, if* $\theta$ *is integral then* $\eta$ *is an algebraic integer.*

*Proof.* If $x_2\theta = B\{\xi\}$ then $\xi$ is a unit of $R(\theta)$ , by Lemma 1 of [9]. Because $A\{\eta\}$ normalizes $\pi K\theta$ ,

$$A\{\eta\}\ B\{\xi\}\ A\{\eta\}^{-1} = \begin{bmatrix} 1-\xi\eta & \xi\eta^2 \\ -\xi & 1+\xi\eta \end{bmatrix} \in \pi K\theta \ .$$

Hence the entries of this last matrix belong to $R(\theta)$ , and because $\xi$ is a unit there we get $\eta \in R(\theta)$ . This proves the theorem. □

The result can be restated as a necessary condition on the longitude polynomial $Q$ for $A_\infty$ to contain $A\{(m+ng)/r\}$ for a given allowable triple $m, n, r$ . This condition is not sufficient in the very simplest case of all, viz. the figure-eight knot where $g = 2\sqrt{-3} = 2 + 4\omega$ , $\omega = (-1 + \sqrt{-3})/2$ , cf. [12]. Here $A\{\omega\}$ does not normalize $\pi K\theta$ yet $\langle\pi K\theta , A\{\omega\}\rangle$ is still discrete. But we think that for most knots the EH-translations normalizing $\pi K\theta$ are exactly those allowed by this criterion. G. Burde in [1] discussed a property of the Alexander polynomial of a knot admitting an axis of rotation which might be considered as a partial analogue of Theorem 2. There should also be a restriction on the Alexander polynomial of a knot admitting other symmetries which preserve both space and knot orientation. In particular, the Alexander polynomial $\Delta(t)$ of a 2-bridge knot $(\alpha , \beta)$ where $\beta^2 \equiv +1 (\mathrm{mod}\ 2\alpha)$ ought to have the shape

$$\Delta(-t^2) = D(t) \ . \ D(-t) \ ,$$

for some $D(t) \in \mathbb{Z}[t]$ .

There does not seem to be a way to decide the existence of a normalizing rotation $R_t$ from $g(\theta)$ or $Q$ alone. Perhaps this is related to the difficulty of proving invertibility using the cyclic knot invariants. Nevertheless, the calculations that one makes in deciding the invertibility of an excellent knot from the hyperbolic structure are very similar to those for the other kinds of symmetries. We feel that the lack of an easy algebraic restriction for invertibility does not mean that the invertibility problem is really harder than the other symmetry problems.

The following peculiar result is a simple application of an argument of T. Jørgensen, cf. §4 of [5]. The computational proof here is rather less elegant than Jørgensen's common perpendicular or Lie product proofs.

THEOREM 3. *Let* k *be an excellent knot such that* $\pi K$ *admits an orientation preserving automorphism* s *such that, for some meridian* $x \in \pi K$, $\pi K$ *is generated by* $\{s^{\nu}(x) : \nu \in \mathbb{Z}\}$. *Then* k *is invertible.*

*Proof.* There is an excellent p-rep $\theta$ of $\pi K$ normalized so that $x\theta = A$, $s(x)\theta = B\{u^2\}$ for some $u \in \mathbb{C}$. Let $S \in P\Gamma L(\mathbb{C})$ correspond to s by $s(y)\theta = Sy\theta S^{-1}$ for $y \in \pi K$, then because s is orientation preserving we have $S \in PSL(\mathbb{C})$. An easy matrix calculation shows that S has the shape

$$S = \begin{bmatrix} 0 & -u^{-1} \\ u & \alpha \end{bmatrix}, \quad \text{where } \alpha \in \mathbb{C},$$

as we noted in §4 of [14]. Write

$$R = \begin{bmatrix} i & i/u \\ 0 & -i \end{bmatrix},$$

then another easy calculation shows that $RS = S^{-1}R$. Hence

$$RS^{\nu}AS^{-\nu}R = S^{-\nu}RARS^{\nu} = S^{-\nu}A^{-1}S^{\nu} \in \pi K\theta,$$

because S normalizes $\pi K\theta$. But the elements $S^{\nu}AS^{-\nu}$ generate $\pi K\theta$, so R normalizes $\pi K\theta$ and Theorem 3 is proved. □

This theorem makes it seem unlikely that an excellent non-invertible n-bridge knot can ever admit an axis of symmetry of order $\geq n$.

## 2. The formalism for 2-bridge knots

Each pair of relatively prime odd integers $\alpha, \beta$ with $\alpha \geq 3$ corresponds to a knot $k = (\alpha, \beta)$ in H. Schubert's normal form, cf. [17]. This knot has a regular projection so that, after orienting k, we read off the standard presentation for $\pi K$, viz.

$$\pi K = |x_1, x_2 : wx_1 = x_2 w : \gamma = w^{-1}\tilde{w}x_1^{2\sigma}| . \tag{2.1}$$

Here $x_1, x_2$ are oriented meridians corresponding to the two bridges, $w = w\{x_1, x_2\}$ is a particular word on $x_1, x_2$ of the shape

$$w = x_1^{\varepsilon_1} x_2^{\varepsilon_2} x_1^{\varepsilon_3} \cdots x_2^{\varepsilon_{\alpha-1}},$$

and the exponents $\varepsilon_j$ are determined by a certain rule from the knot parameters $\alpha$ , $\beta$ , cf. [9]. The sole relation needed to present $\pi K$ on $x_1$ , $x_2$ is $wx_1 = x_2 w$ . Also $\gamma$ is an oriented longitude of $\pi K$ such that $\langle x_1 , \gamma\rangle$ is the standard peripheral subgroup of the previous section. Let

$$\tilde{w} = \tilde{w}\{x_1 , x_2\} = w\{x_1^{-1} , x_2^{-1}\} = x_1^{-\varepsilon_1} x_2^{-\varepsilon_2} x_1^{-\varepsilon_3} \dots x_2^{-\varepsilon_{\alpha-1}}$$

and

$$\sigma = \sum_{j=1}^{\alpha-1} \varepsilon_j .$$

Then an expression for $\gamma$ as a word on $x_1$ , $x_2$ is the final item of (2.1). From the normal form $(\alpha , \beta)$ one can show that $\tilde{w}x_1 = x_2\tilde{w}$ also holds in $\pi K$ .

Every p-rep of our 2-bridge knot group $\pi K$ is simply equivalent to a unique normalized p-rep $\theta = \theta\{\xi\}$ where $x_1\theta = A$ , $x_2\theta = B\{\xi\}$ for some $\xi \in \mathbb{C}$ , and all p-reps mentioned are presumed to be in this normal form. To find them let $\mathbb{Z}[y]$ be the domain of integral polynomials in one variable and consider the elements

$$B = B\{y\} , \; W = w\{A , B\} = A^{\varepsilon_1} B^{\varepsilon_2} \dots B^{\varepsilon_{\alpha-1}} = \begin{bmatrix} w_{11}(y) & w_{12}(y) \\ w_{21}(y) & w_{22}(y) \end{bmatrix}$$

of $PSL(\mathbb{Z}[y])$ . We showed in [9] that $w_{21}(y) = -yw_{12}(y)$ , whence $W$ and $\tilde{W} = \tilde{w}\{A , B\} = w\{A^{-1} , b^{-1}\}$ actually have the shape

$$W\{y\} = \begin{bmatrix} w_{11} & w_{12} \\ -yw_{12} & w_{22} \end{bmatrix} , \; \tilde{W}\{y\} = \begin{bmatrix} w_{11} & -w_{12} \\ yw_{12} & w_{22} \end{bmatrix} = R_0 W R_0 .$$

The matrix equation $WA = BW$ unwinds to four equations on the entries $w_{ij}(y)$ , but two of these are redundant because of the overall setup $(\det A = \det B , \operatorname{tr} A = \operatorname{tr}(B))$ , and a third is redundant because of special properties of the exponent sequence $\vec{\varepsilon}$ , cf. [9]. Only one equation, $\Lambda(y) := w_{11}(y) = 0$ remains as the p-rep condition. Therefore for $\xi \in \mathbb{C}$ the assignment $\theta = \theta\{\xi\}$ as above is a p-rep if and only if $\Lambda(\xi) = 0$ . We showed in [9] that $\Lambda$ has the shape

$$\Lambda(y) = 1 + c_1 y + c_2 y^2 + \dots + c_{\lambda-1} y^{\lambda-1} + (-y)^{\lambda} \ , \quad \text{where} \quad \lambda = \frac{\alpha-1}{2} ,$$

and in [9,11] we showed that the discriminant of $\Lambda$, $\delta\Lambda$, is an odd integer. It followed that $\Lambda$ has no repeated factors.

When $y = \xi$ where $w_{11}(\xi) = 0$ the matrices $W, \tilde{W}$ become

$$W = \begin{bmatrix} 0 & w_{12} \\ -w_{12}^{-1} & w_{22} \end{bmatrix} , \quad \tilde{W} = \begin{bmatrix} 0 & -w_{12} \\ w_{12}^{-1} & w_{22} \end{bmatrix} ,$$

because $\det W = 1$. Hence

$$\gamma\theta = \begin{bmatrix} -1 & -2(w_{12}w_{22} + \sigma) \\ 0 & -1 \end{bmatrix} = A\{g\} \ , \quad \text{so} \quad g = 2(w_{12}w_{22} + \sigma) \ .$$

(These $-1$'s on diag $\gamma\theta$ mean we have to change sign. Our computer printout does not always make it clear whether the signs were changed an even or an odd number of times, so all quoted values for $g$ in the following sections are at least slightly uncertain). Hence the longitude parameter $g = g(\xi)$ of $\theta\{\xi\}$ has the shape $g = 2g_0$ where $g_0 \in \mathbb{Z}[\xi] = R(\theta)$. Then $A\{g_0\}$ normalizes $\pi K \theta$ because $A\{g_0\} \rightleftarrows A$, and a formal calculation shows that

$$A\{g_0\}\ B\{\xi\}\ A\{g_0\}^{-1} = \begin{bmatrix} 1 - w_{12}^{-1}w_{22} & w_{22} \\ -\xi & 1 + w_{12}^{-1}w_{22} \end{bmatrix} = W^{-1}AW \ .$$

Therefore both $R_0$ and $A\{g_0\}$ generate automorphisms of $\pi K \theta$ by conjugation. Because the discriminant $\delta\Lambda$ is odd the prime 2 is unramified in the algebraic number field $\mathbb{Q}(\xi) = F(\theta)$, and because $2 = i(1 - i)^2$ is ramified in $\mathbb{Q}(i)$ we cannot have $i \in f(\theta)$. In particular, $R_0$ cannot belong to $\pi K \theta$, so conjugation by $R_0$ is always an outer automorphisms of $\pi K \theta$. We calculate that

$$WA\{g_0\} = \begin{bmatrix} 0 & w_{12} \\ -w_{12}^{-1} & 0 \end{bmatrix} , \tag{2.2}$$

and this is an involution because the trace is zero. Therefore if $\pi K\theta$ is torsion-free then $A\{g_0\}$ cannot belong to $\pi K \theta$, and this is the situation for excellent p-reps. (It often happens that $A\{g_0\} \notin \pi K \theta$ even when $\theta$ has torsion. Indeed, the only times that we found $A\{g_0\} \in \pi K \theta$ were when $A\{g\} = A^{2n}$ for some non-zero integer $n$, so that $(\pi K \theta)_\infty$ was cyclic.) When $A\{g_0\} \notin \pi K \theta$ it induces an outer automorphism that differs from that of $R_0$. Write $a, r$ for the respective images of these automorphisms in $\mathrm{Out}(\pi K \theta)$, then it is easy to see that either $a = E$, so $\langle a, r\rangle$ is cyclic of order 2, or $\langle a, r\rangle$ is the Klein 4-group. Write

$$D_n = |c, d : c^2 = d^2 = (cd)^n = E|$$

as standard notation for the dihedral group of order $2n$, then $\langle a, r\rangle$ is isomorphic to either $D_1$ or $D_2$.

## 3. Three 2-bridge knots

We now come to the explicit results for $5_2$, $7_7$ and $6_3$. For each we give the excellent factor $\Lambda_1$ of the p-rep polynomial, and an approximation to the excellent root $\xi$ of $\Lambda_1$ which defines $\xi$ completely as the limit of Newton's iteration starting at the quoted value. We also give an approximation to the longitude entry $g(\theta)$, and we shall either give the longitude polynomial $Q(y)$ itself, or else $Q_h(y)$, the primitive irreducible polynomial annihilating $g(\theta)/2$. We then describe the Ford domain $D = D(\pi K \theta)$ by listing the non EH-elements of $\pi K \theta$ that the computer chose to be side pairing transformations for $D$. For each such chosen generator $T$ the isometric spheres $I(T)$, $I(T^{-1})$ are H-planes carrying sides of $D$, and $T$ maps the side on $I(T)$ onto that on $I(T^{-1})$. We also give some numerical data for $I(T^{\pm 1})$ derived from the matrix for $T$ which is useful in interpreting the diagram of $D$ and for seeing how the elements of $A_\infty$ in the normalizer $A \subset P\Gamma L(\pi K \theta)$ act as automorphisms of $\pi K \theta$. The diagram of $D$ (and the later diagram for $D(A)$) is not, strictly speaking, essential for verifying that what the computer did was a correct proof of the stated conclusions, but until these calculations have become a familiar routine it is important for understanding it. Then we quote the cycle transformations that the computer found corresponding to the cycles of non EH-edges of D. The Poincaré presentation of $\pi K \theta$ is formed by setting the cycle transformations to $E$ (the identity) and adjoining the relation $A \rightleftarrows A\{g\}$. We then give a similar account for $A$, its Ford domain $D(A)$, and we state the Poincaré presentation exactly because some of the non EH-cycles are torsion cycles.

The diagrams for the Ford domain $D$ can be used as analogue computers of very low accuracy, and the entire calculation is based on floating point arithmetic of greater but obviously limited accuracy. The conclusions that we derive are absolutely exact and are rigorously proved, at least for the 2-bridge knots. Our method for doing this is based on Poincaré's Theorem for Fundamental Polyhedra that H. Seifert proved rigorously and in generality in [18]. Our method is explained in detail in [16], and the essential secret why it worked so well for the examples below is that for each of these $D(\pi K \theta)$, every non EH-cycle transformation contains exactly 3 non EH-generators. This makes it automatic that $T$ maps the side of $D$ on $I(T)$ exactly, instead of merely approximately, on the side on $I(T^{-1})$, and it was the apparent difficulty of proving just this that made it seem that Poincaré's Theorem was of very limited utility. Once we prove that $\pi K \theta$ is discrete the setup guarantees that $A$ is also discrete and that the maximal isometric spheres for both groups are the same. Then only very low accuracy arithmetic (such as the implied accuracy of our diagrams as reproduced here) is needed to sort out how the non EH-edges are mapped, and this is all we really do.

The computer did absolutely all of the checking needed to make the 2-bridge calculations rigorous. The main reason why these calculations are rigorous and the 3-bridge calculations less satisfactory is that it was reasonable to program the computer to check that every cycle relation of $\pi K \theta$ is a consequence of the relation of $\pi K$ itself, i.e. that $\theta$ is an isomorphism. (It didn't seem so reasonable for the 3-bridge knots, cf. the next section.)

The interpretation of the diagrams is the same as in [12, 15, 16], but there is one minor complication for groups containing EH-rotations. An asterisk in a diagram marks the fixed point, $ax_0(R_t)$, of a rotation $R_t$ in the group. In some cases $ax_0(R_t)$ coincides with the centre, $cn(T)$, of an isometric sphere $I(T)$. Then only one half of the H-polygon for this sphere is a side of $D(A)$, but the diagram does not indicate which half. The computer took this into account in working out the presentation by an elaborate system of marking the edges of the polygon as "live" or "ghost", and arranging that only the live edges are used in the cycles. The ghost edges still appear in the pictures, and only a detailed analysis of the presentation coupled with knowledge of the rules for choosing live edges would determine the precise Ford domain that the computer selected. But we didn't bother to do this because it seems pointless.

Recall our standard notation: for

$$T = \begin{bmatrix} a & b \\ c & d \end{bmatrix}, \quad ad - bc = 1 , \quad c \neq 0 ,$$

we have $cn(T) = -d/c$ , $cn(T^{-1}) = a/c$ , $rd(T) = 1/|c|$ , for the centres of $I(T)$ , $I(T^{-1})$ , and the radius of $I(T)$ . Also we pick out the entries of $T$ (now allowing $c = 0$ ) by the functions $a_{11}, b_{12}, c_{21}, d_{22}$ , e.g. $c_{21}(T) = c$ , and similarly for the other entries. We write polynomials as truncated power series, thus

$$\{c_0 \; c_1 \; c_2 \; \ldots \; c_n\} := c_0 + c_1 y + c_2 y^2 + \ldots + c_n y^n .$$

Finally, all decimal numbers quoted were rounded to the stated values from much more accurate calculations.

3(a) The knot $5_2$

We use the 2-bridge normal form (7 , 3) for this knot. Then the p-rep polynomial $\Lambda$ is

$$\{1 \;\; -2 \;\; 1 \;\; -1\} ,$$

which is irreducible. The excellent root $\xi$ (with positive imaginary part) of $\Lambda$ and the corresponding longitude entry are

$$\xi \approx 0.21508 + 1.30714i , \quad g \approx -2.49024 + 2.97945i .$$

The polynomial $Q_h(y)$ for $g/2$ is

$$\{17 \;\; 15 \;\; 7 \;\; 1\} .$$

The computer choose $D = D(\pi K \theta)$ so that its non EH-side-pairing transformations are the following words in $A$ , $B = B\{\xi\}$ , and their inverses.

| | cn(T) | $cn(T^{-1})$ | rd(T) |
|---|---|---|---|
| $T_1 = B$ | 0.123 - 0.745i | -0.123 + 0.745i | 0.755 |
| $T_3 = A^{-1}B^{-1}AB$ | -0.417 - 0.927i | -0.338 - 0.562i | 0.570 |
| $T_5 = A^{-1}B^{-1}ABA^{-1}B^{-1}A$ | -0.245 + 1.490i | 0 | 0.869 |
| $T_7 = ABA^{-1}B^{-1}$ | $-cn(T_3)$ | $-cn(T_3^{-1})$ | $rd(T_3)$ |

(The computer used the missing subscripts for the inverses of the chosen generators, and we are afraid that reindexing might introduce clerical errors.) The word $T_5$ here is WA . The computer used $A_1 := A\{g+2\} = A^2A\{g\}$ as its standard second EH-generator of $\pi K\theta$ because $\mathrm{Real}|(g+2)| < 0.5$ . Then the computer found that D has 39 non EH-edges (cf. Figure 1(a)), and arranged these into 8 edge cycles. The cycle transformations for these cycles are as follows.

| cycle | | cycle | |
|---|---|---|---|
| 1 | $AT_1^{-1}A^{-1}T_7T_1$ | 2 | $T_3^{-1}A^{-1}T_1^{-1}AT_1$ |
| 3 | $AT_3^{-1}T_5A^{-1}T_1$ | 4 | $T_3A_1^{-1}A^{-1}T_5^{-1}AT_1$ |
| 5 | $A_1^{-1}T_5^{-1}AT_5T_1$ | 6 | $A_1^{-1}A^{-1}T_5^{-1}T_7AT_1$ |
| 7 | $T_5A^{-1}T_5^{-1}T_1$ | 8 | $AT_5A^{-1}T_7^{-1}T_1$ |

To get the Poincaré presentation just set these words to E and adjoin $A \not\approx A_1$ . The computer checked that these relations are equivalent to the assertion that the p-rep $\theta(\xi)$ is an isomorphism. Thus relation 2 expresses $T_3$ as a word in A , $B = T_1$ , relation 3 similarly expresses $T_5$ , and relation 7 corresponds to the knot group relation $wx_1 = x_2w$ . The closing trick of [16] will apply to $\pi K\theta$ provided only that each of the above transformations has exactly 3 non EH-factors, which is true. Therefore $\theta$ is discrete and faithful, i.e. excellent, D is a Ford domain for $\pi K\theta$ , and the orbit space $U_3/\pi K\theta$ is $S^3 - k$ where k is $5_2$ .

It should be obvious from Figure 1(a) that only the two standard elements $R_0$ and $A\{g/2\}$ are needed to generate $\mathcal{A}$ , starting from $\pi K\theta$ . Write $A_0 = A\{g_1/2\}$ , $R = R_0$ , and $\mathcal{A} = \langle \pi K\theta , A_0 , R\rangle$ . The computer choose $D(\mathcal{A})$ so that its spherical sides lie on the following isometric spheres I(U) .

| | $cn(U) = cn(U^{-1})$ | rd(U) |
|---|---|---|
| $U_1 = RT_1$ | 0.123 - 0.745i | 0.755 |
| $U_2 = T_3RAA_0$ | -0.338 - 0.562i | 0.570 |
| $U_3 = T_5A_0$ | 0 | 0.869 |

The computer then read off the following 8 cycle relations from $D(A)$ , and we adjoin the EH-relations to get the complete Poincaré presentation of $A$ .

$$R A_0 A^{-1} U_1 R A_0 A U_2 R A_0 A U_1 = U_2 U_3 A^{-1} U_1 = U_3 R A U_3 U_1 = E ,$$

$$(U_1 R A_0)^2 = (U_3 R)^2 = U_1^2 = U_2^2 = U_3^2 = R^2 = (R A)^2 = (R A_0)^2 = E ,$$

$$A \rightleftarrows A_0 .$$

This presentation for $A \approx Aut(\pi K)$ is obviously quite redundant but we haven't tried to simplify it. It is clear that $Out(\pi K) \approx D_2$ .

Remark that if we had used the other 2-bridge normal form we would have obtained essentially the same Ford domain $D(\pi K \theta)$ , but the interpretation of its side-pairing transformations in terms of the algebraic formalism for $\pi K$ would have been different. For the two versions of $\pi K \theta$ would be isomorphic by an isomorphism that sends the subgroup $(\pi K \theta)_\infty$ of one group to that of the other, and by MRT this isomorphism can be expressed as a conjugation by an EH-translation.

3(b) The knot $7_7$ .

The p-rep polynomial $\Lambda$ for our second example, $K = 7_7$ = 2-bridge knot (21 , 13) is irreducible and factors as $\Lambda = \Lambda_1 \cdot \Lambda_2$ where

$$\Lambda_1 = \{1 \ \ 1 \ \ 3 \ \ 2 \ \ 1\} , \ \Lambda_2 = \{1 \ \ 0 \ \ 0 \ \ 2 \ \ 4 \ \ 3 \ \ 1\} .$$

We shall show that the root

$$\xi \approx -0.95668 + 1.22719i , \text{ with } g(\xi) \approx 1.77019 + 6.51087i ,$$

of $\Lambda_1 = 0$ is excellent, and that all the roots of $\Lambda_2$ are indiscrete.

We begin with $\theta = \theta(\xi)$ . The computer found that the 16 spherical sides of its chosen Ford domain $D = D(\pi K \theta)$ correspond to some rather long words in $A , B$ , e.g.

$$T_5 = AB^{-1} ABA^{-1} B^{-1} AB^{-1} A^{-1} BA^{-1} B^{-1} ABA^{-1} BAB^{-1} A ,$$

and $T_3$ is the even longer word $WA$ in the 2-bridge formalism.

*Figure 1(a) shows the projection on* $\mathbb{C}$ *of* $\mathbb{D}(\pi K\theta)$ . *Each circle is an isometric circle* $I_0(T)$ *where* T *is a non EH-side pairing transformation of* $\mathbb{D}$ , *and the thin line bisecting* $I_0(T)$ *is* $Ref_0(T)$ , *the reflecting line of* T . *The subgroup* $(\pi K\theta)_\infty$ *maps the isometric spheres and circles shown on others not shown. The bold lines are projections of non EH-edges of* $\mathbb{D}$ , *and* $\mathbb{D}_\infty$ , *the associated fundamental domain in* $\mathbb{C}$ *of* $(\pi K\theta)_\infty$ , *is bounded by the outer circuit of bold segments.*

*To interpret the figure the first step is to label each circle* $I_0(T)$ *by its transformation* T , *using the listed data* cn(T) , rd(T) , *and the helpful fact that the circles* $I_0(T)$ , $I_0(T^{-1})$ *have the same radius and parallel reflecting lines. The side of* $\mathbb{D}$ *on* I(T) *projects to a polygon bounded by bold lines inside* $I_0(T^{-1})$ *which is congruent to the corresponding polygon for* $I(T^{-1})$ *after flipping it about* $Ref_0(T^{-1})$ . *When the polygons are labelled the effect of the transformations on the edges can be read off easily and the edge cycles worked out. Beware that you are likely to get a cyclic permutation of the cycle listed here, perhaps in reversed order. This depends on the starting edge and direction of travel along the cycle.*

Figure 1(a). The Ford domain

$\mathbb{D}(\pi K\theta)$ , $K = 5_2$ .

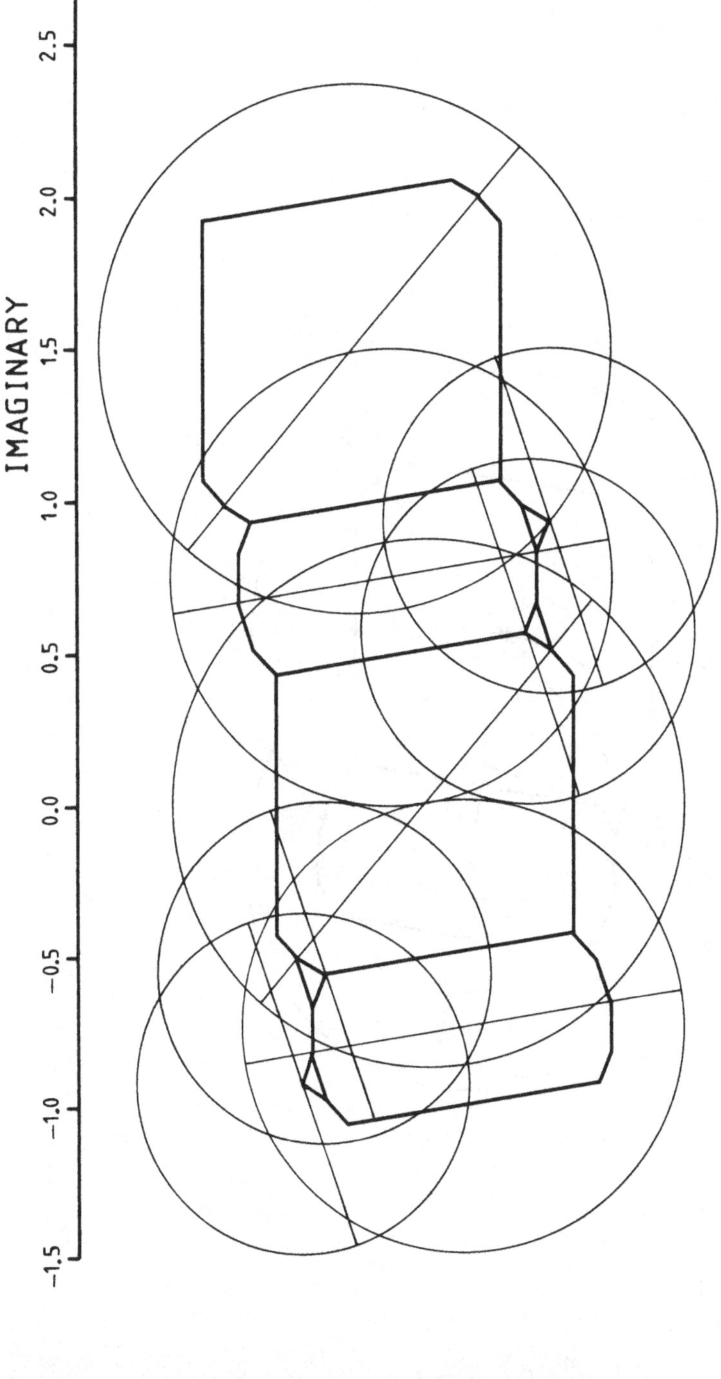

Figure 1(b). The Ford domain for the automorphism group of $\pi K\theta$, $K = 5_2$.

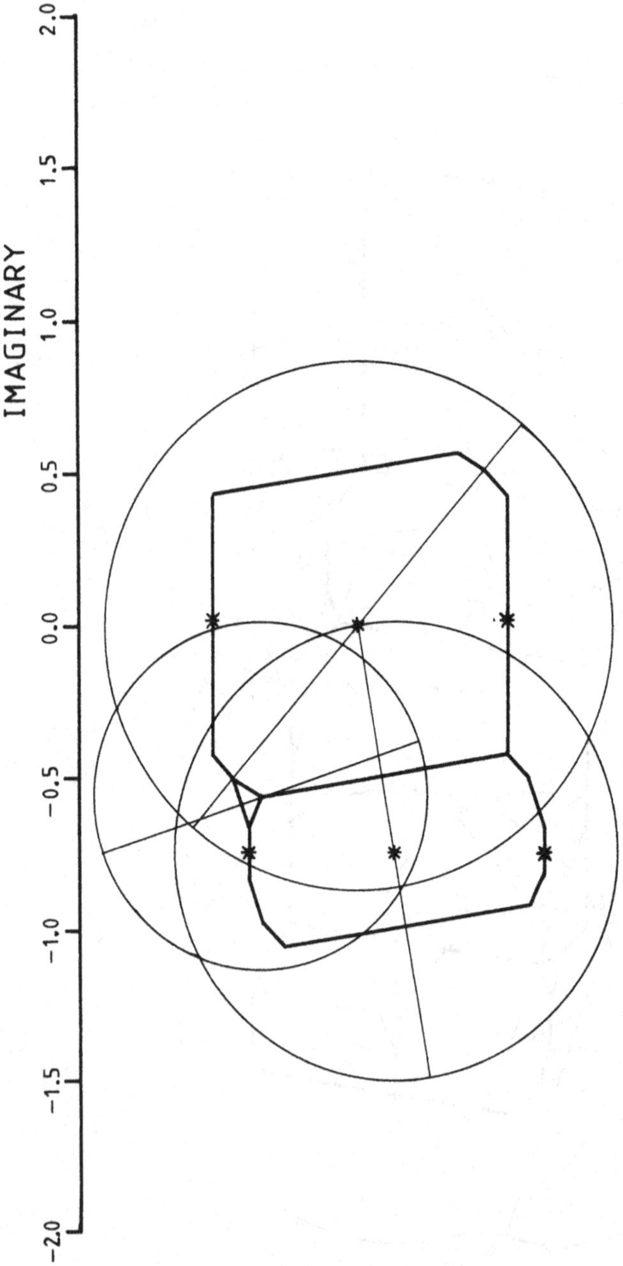

Hence we shall not write out the expressions for all of the $T_j$ used as side-pairing transformations of $D$, but they can easily be deduced from the Poincaré presentation (and were, as part of the check that $\theta$ is an isomorphism). The data for the isometric spheres of these elements is as follows.

| | cn(T) | $cn(T^{-1})$ | rd(T) |
|---|---|---|---|
| $T_1 = B$ | -0.395 - 0.507i | 0.395 + 0.507i | 0.643 |
| $T_3$ | -0.115 + 3.255i | 0 | 0.802 |
| $T_5$ | -0.086 + 2.442i | 0.029 - 0.814i | 0.653 |
| $T_7 = BA^{-1}B^{-1}AB^{-1}A^{-1}B$ | 0.057 - 1.628i | -0.057 + 1.628i | 0.802 |
| $T_9$ | 0.338 + 2.135i | -0.453 + 1.121i | 0.643 |
| $T_{11} = R_0T_5R_0$ | $-cn(T_5)$ | $-cn(T_5^{-1})$ | $rd(T_5)$ |
| $T_{13} = R_0T_9R_0$ | $-cn(T_9)$ | $-cn(T_9^{-1})$ | $rd(T_9)$ |
| $T_{15}$ | 0.490 + 2.749i | -0.490 - 2.749i | 0.643 |

The computer used $A_1 := A^{-2}A\{g\}$ as its second EH-generator and found that the 72 non EH-edges of $D$ lie in 16 edge cycles whose cycle transformations are as follows.

| cycle | | cycle | |
|---|---|---|---|
| 1 | $T_3\ A^{-1}\ T_3^{-1}T_1$ | 2 | $A^{-1}\ T_3\ A_1\ T_{11}^{-1}\ T_1$ |
| 3 | $T_5\ T_3^{-1}\ A^{-1}\ T_1$ | 4 | $A^{-1}\ T_5\ A^{-1}\ T_9^{-1}\ A^{-1}\ T_1$ |
| 5 | $A^{-1}\ T_{13}\ A^{-1}\ T_{11}^{-1}\ A^{-1}\ T_1$ | 6 | $T_{15}^{-1}\ A^{-1}\ T_{11}^{-1}\ T_3$ |
| 7 | $A^{-1}\ T_{15}^{-1}\ A_1^{-1}\ A^{-1}\ T_3^{-1}\ A\ T_3$ | 8 | $A_1^{-1}\ T_{15}\ A\ T_5^{-1}\ T_3$ |
| 9 | $T_7\ A\ T_{13}^{-1}\ T_5$ | 10 | $T_9^{-1}\ A^{-1}\ T_7\ T_5$ |
| 11 | $T_{15}^{-1}\ T_{13}^{-1}\ A\ T_5$ | 12 | $A\ T_7^{-1}\ T_9\ T_7$ |
| 13 | $T_{11}^{-1}\ T_9\ A\ T_7$ | 14 | $T_{13}^{-1}\ T_7^{-1}\ A\ T_7$ |
| 15 | $A^{-1}\ T_{13}\ T_{11}^{-1}\ T_7$ | 16 | $T_{15}^{-1}\ T_{11}^{-1}\ A\ T_9$ |

The Poincaré presentation is derived as usual, and the computer checked everything. This completes the proof that $\xi$ is excellent.

This time inspection of the diagram, Figure 2(a), of $D$ suggests that $A_0 := A\{g_0\}$ where $g_0 = (g - 2)/4$ should normalize $\pi K \theta$. The polynomial $Q_h$ and the primitive irreducible polynomial $Q_4$ annihilating $g_0$ are

$$Q_h = \{83 \quad 30 \quad 12 \quad 2 \quad 1\}, \; Q_4 = \{8 \quad 8 \quad 6 \quad 3 \quad 1\}.$$

We calculate that

$$g_0 = -1 + 2\xi + \xi^2 + \xi^3,$$

and that

$$A_0 B A_0^{-1} = \begin{bmatrix} 1 + g_0 & 3 + 2\xi + 3\xi^2 + \xi^3 \\ -\xi & 1 - g_0 \end{bmatrix} = T_9 \in \pi K \theta .$$

Such calculations are calculations with polynomials in $\mathbb{Z}[y]$ taken modulo $\Lambda_1(y)$, and are formal manipulations with rows of integers. Hence they are rigorous proofs of these assertions, and we conclude that $A_0$ indeed normalizes $\pi K \theta$. Write $R = R_0$ again, and let $A := <\pi K \theta, A_0, R>$. Then $A$ is a discrete group containing $\pi K \theta$ as a normal subgroup of index 8. The computer choose the Ford domain $D(A)$ so that its non EH-side-pairing transformations are the following.

| | $cn(U) = cn(U^{-1})$ | $rd(U)$ |
|---|---|---|
| $U_1 = RT_1$ | $-0.395 - 0.507i$ | 0.643 |
| $U_2 = T_3 A_0^2$ | 0 | 0.802 |
| $U_3 = T_5 A_0^2$ | $0.029 - 0.814i$ | 0.653 |

Then the complete Poincaré presentation of $A$ deduced from $D(A)$ is :

$$R A_0 A U_1 R A_0 A U_3 A U_1 = U_2 R A U_2 U_1 = R A U_2 U_3 U_1 = E,$$

$$(U_2 R)^2 = (U_3 R A_0)^2 = U_1^2 = U_2^2 = U_3^2 = R^2 = (R A)^2 = (R A_0)^2 = E,$$

$$A \overset{\longleftarrow}{\longrightarrow} A_0 .$$

The definition of $A_0$ shows that $A_0^2 \notin \pi K \theta$, $A_0^4 \in \pi K \theta$. Hence $A_0$ represents an element of order 4 in Out $(\pi K \theta)$, whence

Figure 2(a). The Ford domain

$\mathbb{D}(\pi K\theta)$, $K = 7_7$.

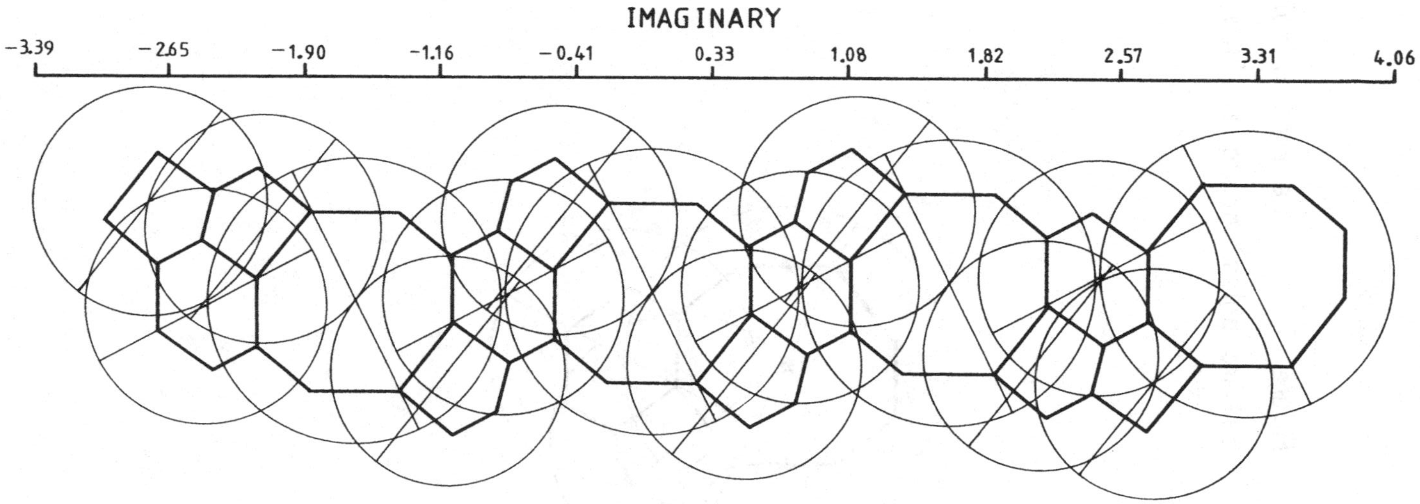

Figure 2(b). The Ford domain for the automorphism group of $\pi K \theta$, $K = 7_7$.

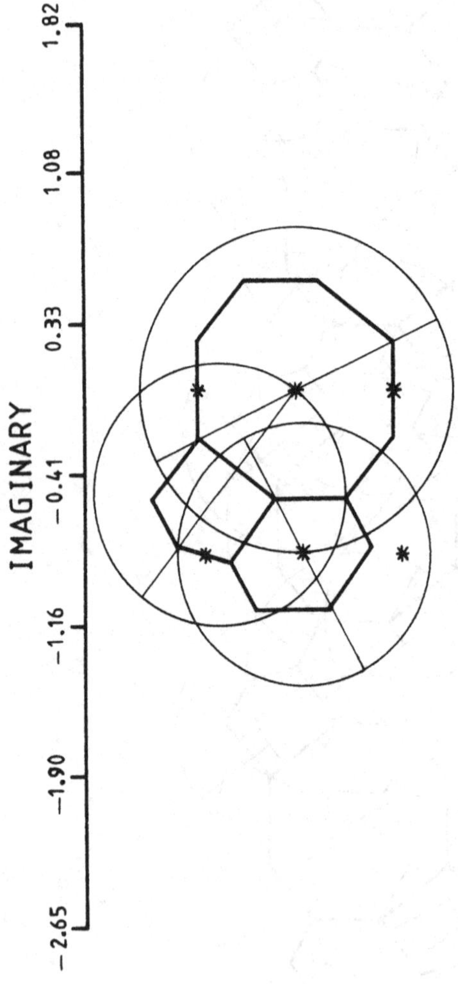

Out $(\pi K \theta) \approx D_4$ .

Our last item for $7_7$ is the proof that all the roots of $\Lambda_2 = 0$ give rise to p-reps on indiscrete groups. Our proofs of indiscreteness are always based on Shimizu's Lemma, which asserts that if $G = \langle A, T\rangle \subset SL(\mathbb{C})$ then $G$ is indiscrete when $0 < |c_{21}(T)| < 1$ . For a proof see p.58 of [6]. The roots of $\Lambda_2 = 0$ are

$$\xi_1 \approx -0.75377 + 0.99896i \ , \ \xi_2 \approx -1.16236 + 0.63545i \ ,$$

$$\xi_3 \approx 0.41613 + 0.43668i \ ,$$

and their complex conjugates. Because $0 < |\xi_3| < 1$ , Shimizu's Lemma applies directly to prove $\langle A, B\{\xi_3\}\rangle$ is indiscrete. Write $B = B\{y\} \in PSL(\mathbb{Z}[y])$ , and let

$$W_1 = A^{-1}\ B\ A\ B^{-1}\ , \ V = W_1^{-1}\ B\ W_1\ A^{-1} \in PSL(\mathbb{Z}[y]) \ .$$

We calculate that

$$c_{21}(V) = \{0 \quad 1 \quad 2 \quad 3 \quad 2 \quad 1\} \ ,$$

and that $c_{21}(V)(\xi_1) \approx -0.01753 - 0.36344i$ , which has absolute value $< 1$ . Hence $\langle A, V\{\xi_1\}\rangle \subset \langle A, B\{\xi_1\}\rangle$ is already indiscrete. Finally, for $\xi_2$ take $B = B\{y\}$ ,

$$V = BA^{-1}\ B^{-1}\ AB^{-1}\ A^{-1} = \begin{bmatrix} 1 - y - y^2 & -1 + y^2 \\ y(1+y)^2 & 1 - y^2 - y^3 \end{bmatrix} ,$$

and compute that $c_{21}(V)(\xi_2) \approx 0.56984$ (a real number). Hence the last of the roots of $\Lambda_2$ is indiscrete, and this proves all our assertions for $7_7$ . Incidentally, the polynomial $Q_h$ for $\Lambda_2$ is

$$Q_h = \{-7 \quad -1 \quad -1 \quad 1\} \ ,$$

so $\mathbb{Q}(g(\theta))$ is a proper subfield of $F(\theta)$ for these p-reps.

### 3(c) The knot $6_3$

The knot $K = 6_3$ has the 2-bridge normal form (13,5) and its p-rep polynomial,

$$\Lambda = \{1 \quad -1 \quad 2 \quad -4 \quad 5 \quad -3 \quad 1\}$$

is irreducible. The excellent root and corresponding longitude entry are

$$\xi \approx 0.84116 + 1.20014i \ , \ g(\xi) \approx 5.51057i \ .$$

Some of the side-pairing transformations for the Ford domain $D(\pi K \theta)$ are

$$T_7 = AB^{-1}A^{-1}BABA^{-1}B^{-1}A \ , \ T_5 = ABT_7A^{-1} \ , \ T_3 = A^{-1}B^{-1}T_5 = W \ .$$

The data for the spherical sides of $D$ are:

| | cn(T) | cn($T^{-1}$) | rd(T) |
|---|---|---|---|
| $T_1 = B$ | $0.392 - 0.559i$ | $-0.392 + 0.559i$ | 0.682 |
| $T_3$ | $2.755i$ | 0 | 0.826 |
| $T_5$ | $-0.108 + 1.936i$ | $0.108 + 0.819i$ | 0.682 |
| $T_7$ | $-\frac{1}{2} + 1.378i$ | $\frac{1}{2} - 1.378i$ | 0.826 |
| $T_9 = R_0 T_5 R_0$ | $-\mathrm{cn}(T_5)$ | $-\mathrm{cn}(T_5^{-1})$ | $\mathrm{rd}(T_5)$ |
| $T_{11} = T_9^{-1} T_3 A^{-1}$ | $0.392 + 2.197i$ | $-0.392 - 2.197i$ | 0.682 |

This time the computer used $A_1 = A\{g\}$ itself as the second EH-generator. The domain $D$ has 56 non EH-edges arranged into 12 edge cycles whose cycle transformations are the following.

| Cycle | | cycle | |
|---|---|---|---|
| 1 | $T_3 A^{-1} T_3^{-1} T_1$ | 2 | $A T_3 T_5^{-1} T_1$ |
| 3 | $T_7 A^{-1} T_5^{-1} A T_1$ | 4 | $T_9 A_1^{-1} T_3^{-1} A T_1$ |
| 5 | $A T_9 A^{-1} T_7 T_1$ | 6 | $T_{11}^{-1} A_1^{-1} T_3^{-1} A T_3$ |
| 7 | $A^{-1} T_{11}^{-1} T_9^{-1} T_3$ | 8 | $A_1 A T_{11} T_5^{-1} T_3$ |
| 9 | $T_7^{-1} A^{-1} T_7 A^{-1} T_5$ | 10 | $A^{-1} T_{11}^{-1} A^{-1} T_7 T_5$ |
| 11 | $A T_7^{-1} A T_9 T_7$ | 12 | $A^{-1} T_{11}^{-1} A^{-1} T_9^{-1} T_7$ |

It might have been difficult to derive the expressions for the $T_j$ as words in $A, B$ from the Poincaré presentation of $\pi K \theta$ if we had only known $T_1 = B$ in advance. The computer checked everything, so $\theta$ is excellent.

The discussion of $\mathrm{Aut}(\pi K \theta)$ is naturally split into two parts, the first being of $A^+ \approx \mathrm{Aut}^+(\pi K \theta)$, and the second of the effect of the glide-reflection generating $A \approx \mathrm{Aut}(\pi K \theta)$ from $A^+$. The splitting was reinforced by the fact that our programs do not (yet?) take account of the orientation-reversing transformations in the necessary generality, so we had to do the work for $A$ by hand.

The diagram, Figure 3(a), for $D$ suggests that the only elements of $A^+_\infty$ needed to generate $A^+$ from $\pi K \theta$ are the standard normalizers, $A_0 = A\{g/2\}$ and $R = R_0$. The computer choose $D(A^+)$ so that its spherical sides are as follows.

| | $cn(U) = cn(U^{-1})$ | $rd(U)$ |
|---|---|---|
| $U_1 = RT_1$ | $0.392 - 0.559i$ | $0.682$ |
| $U_2 = T_3 A_0$ | $0$ | $0.826$ |
| $U_3 = RA_0 T_9^{-1}$ | $-0.108 - 0.819i$ | $0.682$ |
| $U_4 = A_0^{-1} T_7 A$ | $-\frac{1}{2} - 1.378i$ | $0.826$ |

The Poincaré presentation for $A^+$ was then worked out and found to be:

$$U_2 RAU_2 U_1 = AU_2 U_3 U_1 = AU_3 AU_4 A^{-1} U_1 = U_4 RA_0 A^2 U_4 U_3 = E ,$$
$$(U_2R)^2 = (U_4 RA_0A)^2 = U_1^2 = U_2^2 = U_3^2 = U_4^2 = R^2 = (RA)^2 = (RA_0)^2 = E ,$$
$$A \rightleftarrows A_0 .$$

If we had not already known that $6_3$ was amphicheiral the fact that $g$ seems to be pure imaginary would be a tip to look for a glide-reflection $J$ normalizing $\pi K \theta$. Examination of Figure 3(a) suggests that

$$J : z \mapsto -\bar{z} + \frac{2-g}{4}, \text{ with } J^{-1} : z \mapsto -\bar{z} + \frac{2+g}{4} \quad (3.1)$$

should work, and that we should expect to find

$$JAJ^{-1} = A^{-1} , \; JBJ^{-1} = AT_9 = B^{-1} A^{-1} BAB^{-1} A^{-1} B^{-1} AB. \quad (3.2)$$

Figure 3(a). The Ford domain $\mathbb{D}(\pi K\theta)$,

$K = 6_3$.

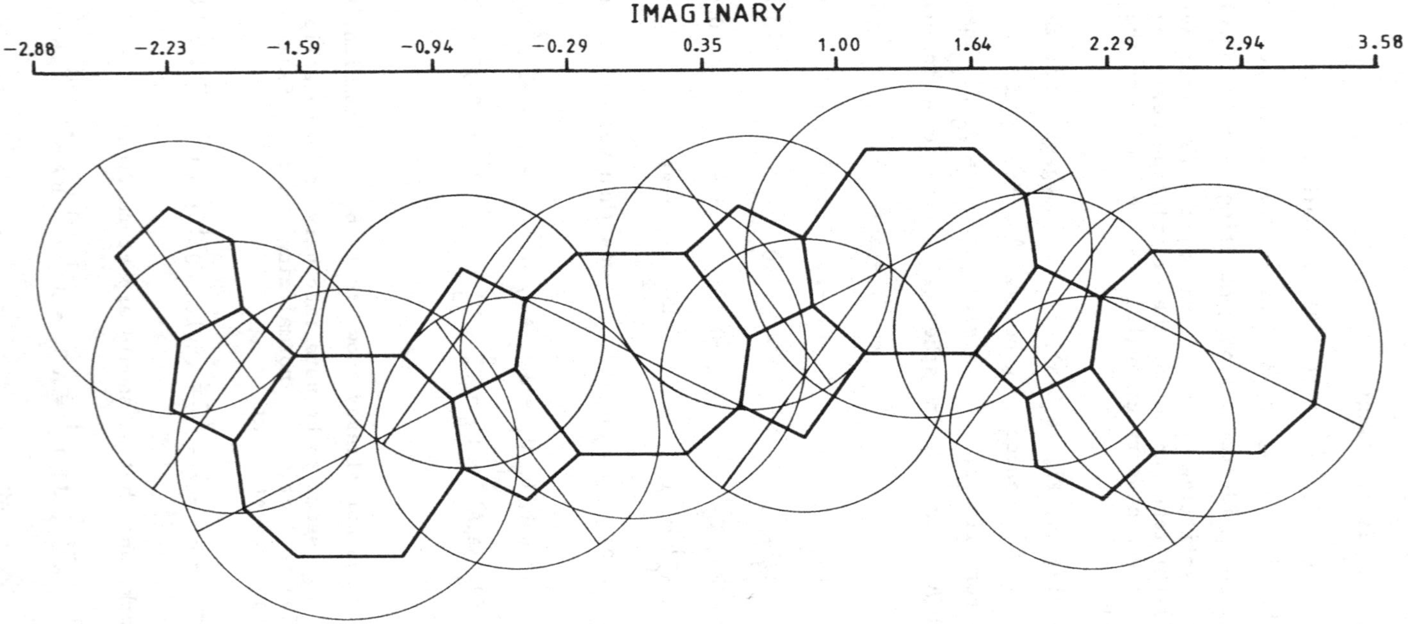

Figure 3(b). A Ford domain for the orientation-preserving subgroup $A^+$ of the automorphism group A of $\pi K\theta$, $K = 6_3$. The shaded region indicates a Ford domain for A itself.

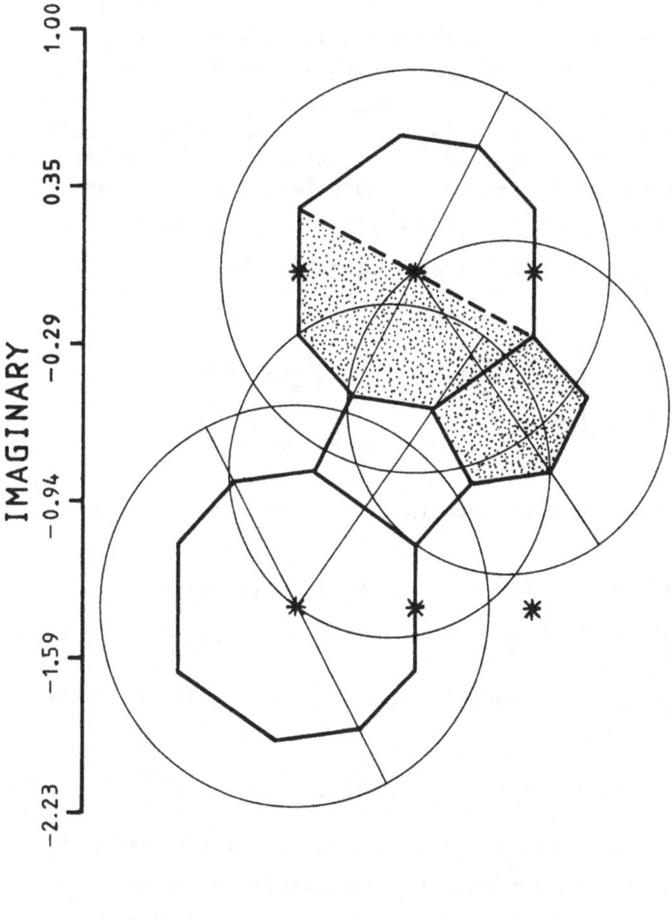

We verified that the $J$ of (3.1) indeed satisfies (3.2) by the method used for $7_7$, viz. calculation in $\mathbb{Z}[y]$ modulo $\Lambda(y)$. Another way to do this will be used in section 5 for an automorphism for $9_{48}$.

First of all, supplementary computer calculations in $\mathbb{Z}[y]$ modulo $\Lambda(y)$ told us that if

$$P = \{-2 \quad 0 \quad -8 \quad 16 \quad -12 \quad 4\} \in \mathbb{Z}[y] \,, \qquad (3.3)$$

then $g(\xi) = P(\xi)$, and that the longitude polynomial is

$$Q = \{6336 \quad 0 \quad -592 \quad 0 \quad 4 \quad 0 \quad 1\} \,.$$

Now $Q$ is even (cf. Theorem 1 with $e = 0$), so each root $g'$ of $Q = 0$ which is neither real nor pure imaginary is one of a quadruple $\{\pm g' , \pm \bar{g}'\}$ of roots. The roots $g'$ for the indiscrete p-reps are roughly $\pm (3.7 \pm 0.8i)$, so there are only two further roots. Thus the approximate value for $g$ quoted above is adequate to prove that $g$ is pure imaginary. Using this one proves $JAJ^{-1} = A^{-1}$ quite easily. The assertion for $JBJ^{-1}$ is rather harder. A formal conjugation by the $J$ of (3.1) gives

$$JBJ^{-1} = \begin{bmatrix} 1 + (\frac{2-g}{4})\,\bar{\xi} & -(\frac{2-g}{4})^2\,\bar{\xi} \\ \bar{\xi} & 1 - (\frac{2-g}{4})\,\bar{\xi} \end{bmatrix}$$

If we write $B = B\{y\}$ and compute in $\mathbb{Z}[y]$ the word $AT_9\{y\}$ of (3.2) expands to

$$\begin{bmatrix} \{1 \quad 0 \quad 1 \quad -1 \quad 1 \quad 0\} & \{0 \quad 0 \quad 0 \quad -1 \quad 0\} \\ \{0 \quad 1 \quad -2 \quad -3 \quad -2 \quad 1\} & \{1 \quad 0 \quad -1 \quad 1 \quad -1\} \end{bmatrix}.$$

Write $f(y) = c_{21}(AT_9\{y\})$, then another longish formal calculation gives

$$\Lambda(f(y)) \equiv 0 (\text{mod } \Lambda(y)) \,.$$

This implies that $\xi_j \mapsto f(\xi_j)$ is a permutation of the roots $\xi_j$ of $\Lambda = 0$, and a calculation with the approximate value of $\xi$ shows that $f(\xi) = \bar{\xi}$. Therefore $c_{21}(JBJ^{-1}) = c_{21}(AT_9)$ for $\pi K \theta$. From (3.3) we get

$$\frac{2-g}{4} = \{1 \quad 0 \quad 2 \quad -4 \quad 3 \quad -1\}\ (\xi)\ .$$

A final formal calculation gives

$$\{1 \quad 0 \quad 2 \quad -4 \quad 3 \quad -1\}\ .\ f(y) \equiv \{0 \quad 1 \quad -1 \quad 1\}\ (\mathrm{mod}\ \Lambda(y))\ .$$

This tells us that $a_{11}(JBJ^{-1}) = a_{11}(AT_9)$ in $R(\theta)$. Because $JBJ^{-1}$ and $AT_9$ are both parabolic with trace +2 their $d_{22}$-entries must also be equal, and then det = 1 implies that the $b_{12}$-entries are equal. This completes a proof that $J$ normalizes $\pi K\theta$ by the rule (3.2).

We are now ready to exhibit a Ford domain $D(A)$ for the group $A \subset P\Gamma L(\mathbb{C})$ generated by $A^+$ and $J$. Let

$$J_1 := ARA_0 J : z \mapsto \bar{z} + \frac{2-g}{4}\ .$$

Because $J$ and $J_1$ normalize $\pi K\theta$ they must map the lattice of maximal isometric spheres of $\pi K\theta$ on itself, and a calculation of very low accuracy, such as inspection of our diagrams, is good enough to see how they do it. This means that proposed relations such as

$$U_3 = J_1^{-1}\ U_1\ J_1\ ,\ U_4 = J_1^{-1}\ U_2\ J_1\ ,$$

can be verified by inspection of Figure 3(b). As $D(A)$ we take the portion of $U_3$ in $D(A^+)$ that projects to the shaded portion of Figure 3(b). The side-pairing transformations for $D(A)$ are $U_1$, $U_2$, $R$, $RA$, and $J_1$. Note that $J_1$ maps the EH-sides above the H-segments on bdry $D(A)$ where $I(U_4)$ meets $I(U_1)$, $I(U_2)$ onto the EH-sides which meet along an EH-axis in $\mathrm{Ref}(U_2)$. Then it is easy to read off the following Poincaré presentation for $A$.

$$U_2\ RAU_2\ U_1 = A\ U_2\ J_1^{-1}\ U_1\ J_1\ U_1 = E\ ,$$

$$U_1^2 = U_2^2 = (U_2\ R)^2 = R^2 = (RA)^2 = J_1\ A^{-1}\ J_1 = E\ .$$

It is also obvious from $D(A)$ that $A$ is a maximal discrete subgroup of $P\Gamma L(\mathbb{C})$, so $A$ is the full automorphism group of $\pi K\theta$.

It only remains to compute $\mathrm{Out}(\pi K\theta)$ . Let $a, j, r$ be the images of $A_0, J, R$ in $\mathrm{Out}(\pi K\theta)$ , then it is easy to see that

$$J^2 = A_0^{-1}, \quad RJR = A_0 A^{-1} J,$$

whence $j^2 = a^{-1} = a$ and $rjr = aj = j^{-1}$ . Therefore

$$\mathrm{Out}(\pi K\theta) = |r, j : r^2 = j^4 = E, rjr = j^{-1}| \approx D_4 .$$

This completes the analysis of the last of our 2-bridge examples.

We remark that the subgroup $\langle \pi K\theta, J\rangle$ of $A$ is expected to be torsion-free and thus have an orbit space of the shape $N - k_1$ , where $N$ is a closed nonorientable 3-manifold and $k_1$ is an orientation-reversing loop in $N$ . Perhaps this would have been the most interesting subgroup of $A$ , but we aren't set up to consider groups with orientation-reversing elements in a systematic manner.

## 4. Conjectures for 2-bridge knots

One thing that an elaborate analysis of a few examples may not do well is determine how typical they are in the class they represent. In the present case perhaps one would not have suspected that the number of non EH-edge cycles of the Ford domain $D(\pi K\theta)$ for a 2-bridge knot $K$ might be a function of the minimal crossing number of the model $k$ of $K$ just because our 3 examples fit a simple formula, but the same formula fits some 5 dozen further examples, including all cases where the crossing number is $\leq 9$ . There are several such nonobvious regularities in our computer printout, and we will discuss them here. But we must caution that the conjectures these regularities suggest do not seem to be of any real importance. Rather they may be corollaries or more indirect consequences of straightforward theorems or even minor little results. The point is that often regularities in computer printout can be traced to obvious trivial features of the setup, and we find it difficult to appraise the value of many of these regularities.

We begin with the relations between the minimal crossing number and the geometric complexity of a Ford domain. Given a 2-bridge knot type $K$ let $\nu(K)$ denote the least crossing number of any regular projection of any model $k$ of $K$ . Also, using the setup of the preceding two sections, we select a Ford domain $D = D(\pi K\theta)$ with the least possible number, $2f$ say, of E-spherical sides, and let $e$ be the number of its non EH-edge cycles. Then we found that

$$e = 2f \tag{4.1}$$

and

$$e \text{ (or } 2f) = 4(\nu - 3) \tag{4.2}$$

holds for all of our more than 60 worked examples. The first assertion (4.1) also holds for the small number of excellent 3-bridge knots that we have investigated, except for those knots admitting an axis of rotational symmetry of order $\geq 3$. The assertion (4.2) seems to be peculiar to 2-bridge knots, and the number $\nu$ is at most distantly related to the geometric complexity of $D$ for 3-bridge knots.

Looking at $D$ more closely, let $h$ be any real number greater than the radius of any isometric sphere of $\pi K \theta$, let $U_3(h)$ be the portion $\{(z, t) \in U_3 : t \leq h\}$ of $U_3$ at E-height $\leq h$ above $\mathbb{C}$, and let $\Pi_h$ be the slice of $U_3$ at height $h$. Then $\pi K \theta$ rolls $U_3(h)$ up to a 3-manifold $M_h$ with boundary $T_h = \Pi_h / (\pi K \theta)_\infty$, and $M_h$ is homeomorphic to the knot exterior $K$. Let $S$ be the closure of the union of the spherical sides of $D$ and let $\bar{S}$ be its image in $M_h$. The EH-projection of the closure of $D \cap \Pi_h$ on $S$ induces a mapping of $T_h$ on $\bar{S}$ so that $M_h$ is the mapping cylinder of this map. We see that $\bar{S}$ is a spine of $M_h$, and $\bar{S}$ is naturally divided up as a cell complex in which each edge (1-cell) is the image of an edge cycle of $D$ and each face (2-cell) is the image of a pair of sides of $D$. The Euler characteristic $\chi(K) = 0$, hence $\chi(\bar{S}) = 0$. Let $v$ be the number of vertices of $\bar{S}$, then $v$ is also the number of cycles of vertices of $D$. Thus (4.1) is equivalent to the assertion $v = f$. The conjecture (4.2) somehow suggests that there is a standard way of drawing a spine of $K$ when $k$ is a minimal crossing model of the type so that we can visualize $K$ as a mapping cylinder. This would go a long way towards making the excellent hyperbolic structure on $S^3 - k$ into something intuitively natural.

Our original expectation was that the determinant $\alpha$ of $k = (\alpha, \beta)$ would be the number having the greatest influence on the geometric complexity of $D$. The relations (4.1), (4.2) show that this is wrong, but there is a more subtle influence of the size of $\alpha$ on $D$. Our data suggest that for each fixed minimal crossing number $\nu$ the magnitude $|g(\theta)|$ of the longitude entry increases with $\alpha$. Furthermore our Calcomp plots of $D(\pi K \theta)$ for the 2-bridge knots of fixed $\nu$ for $\nu \leq 8$

show a steady progression in the sizes and shapes of the E-projections on $\mathbb{C}$ of the spherical sides of $D$. For the smallest $\alpha$ we see mostly very tiny thin triangles together with a few large many sided polygons that are quite irregular. As $\alpha$ increases the smaller polygons grow larger to decrease the range of sizes of the polygons, and they all become not far from regular polygons with at least 4 sides. Therefore it is easier to work with $D$ both numerically and pictorally for the larger $\alpha$, but the complexity of the algebraic calculations does increase with $\alpha$. In fact the size-shape effect for $D$ is so strong that it is difficult for a computer to pursue the calculations for the twist knots $(\alpha, \alpha-2)$ very far. These knots minimize $\alpha$ among the excellent 2-bridge knots for fixed $\nu$, and the smallness of their projected triangles quickly defeats the Calcomp plotter and soon demands numerical accuracy beyond the capacity of the floating point arithmetic for the machine.

A feature of the Ford domains for the 2-bridge knots that is important for our method of calculation is that every observed non EH-edge cycle transformation is the product of exactly 3 non EH-elements and an irrelevant number of EH-transformations. Thus the spines $\bar{S}$ described above all seem to have three 2-cells meeting along each edge. Another very convenient regularity was that the following process always produced a list containing all the side-pairing transformations $T_j$ of $D(\pi K \theta)$. For each segment $V$ of $W = w\{A, B\}$ that begins and ends with $B^{\pm 1}$ take $V$, $R_0 V R_0$, $A_0 V A_0^{-1}$, $A_0 R_0 V R_0 A_0^{-1}$. Normalize each of these by pre and post multiplication by elements of $(\pi K \theta)_\infty$ to bring $cn(T_j^{\pm 1})$ as near as possible to $0$. Our list was the resulting collection of elements of $\pi K \theta$, and the process of generating it and testing the resulting collection of isometric spheres proved to be so fast that we could easily have computed the excellence of many hundreds of examples.

The results for the automorphism groups were absolutely predictable. Consider $k = (\alpha, \beta)$ where $\beta \not\equiv \pm 1$. When $\beta^2 \not\equiv \pm 1 \pmod{2\alpha}$ we always found that all the automorphisms were the standard ones. When $\beta^2 \equiv -1 \pmod{2\alpha}$ we found in addition the glide-reflection $J : x \mapsto -\bar{z} + (g + 2)/4$ as a normalizer of $\pi K \theta$. When $\beta^2 \equiv +1$ we found that the only non-standard normalizer was $A\{(g+2)/4\}$. In both cases $\beta^2 \equiv \pm 1$ we got $Out(\pi K) \approx D_4$. Our three examples illustrated these three cases and show every symptom of being absolutely typical. Of course a glide-reflection $J$ is required when $\beta^2 \equiv -1 \pmod{2\alpha}$ because

H. Schubert proved in [17] that these knots are amphicheiral. We shall presently say a few words about the need for a non-standard normalizer when $\beta^2 \equiv +1 \pmod{2\alpha}$ .

The non-excellent classes of p-reps for the knots $(\alpha, \beta)$ with $\beta \not\equiv \beta^2 \equiv +1 \pmod{2\alpha}$ seem to have certain distinctive properties illustrated by $7_7$ and $7_4$ . In all cases investigated we found that the p-rep polynomial had a non-excellent irreducible factor $\Lambda_2$ whose roots correspond to p-reps $\theta$ such that: (1) the longitude entry $g(\theta)$ lies in a proper subfield of $F(\theta)$ , (2) $\pi K \theta$ is generated by its elliptic elements, and (3) all p-reps for $\Lambda_2$ are indiscrete. We don't predict that the p-rep polynomial will always be the product of just two irreducible polynomials for these knots, but we do predict the existence of at least one factor $\Lambda_2$ with these properties. Incidentally, all irreducible factors of any p-rep polynomial of degree $\leq 2$ and all known such factors of degree 3 have at least one root $\xi$ such that $\langle A, B\{\xi\}\rangle$ is discrete, but the $\Lambda_2$ factor for $7_4 = (15, 11)$ is a quartic polynomial whose groups are all indiscrete. We also make the general prediction that if $\theta$ is a faithful p-rep for an arbitrary knot which is *rigid* (meaning $\theta$ does not belong to a continuous family of p-reps) then $\mathbb{Q}(g(\theta)) = F(\theta)$ . (When k is a product of $\geq 2$ excellent 2-bridge knots we can easily find a faithful p-rep $\theta$ of $\pi K$ with property (1) above, so an assumption like rigidity is trivially necessary. We don't have enough experience with knots in closed manifolds other than $S^3$ to know whether we should have required $k \subset S^3$ .)

The EH-translations normalizing $\pi K \theta$ for an excellent knot k can be used to construct new manifolds whose homology invariants are algebraic invariants of the knot type K by the following process. Let G be the commutator subgroup of $\pi K \theta$ , then G is a Kleinian group of first kind such that $G_\infty = \langle A\{g\}\rangle$ . (In most cases G is not finitely generated, but when it is G is a J-group as defined in [15].) Let $A_0$ be an EH-translation normalizing $\pi K \theta$ such that $A_0 \notin \pi K \theta$ . For $n \in \mathbb{Z}$ let $G_n = \langle G, A_0 A^n\rangle$ . We expect that for most or all nonzero n the group $G_n$ is torsion-free, so that $G_n$ is the fundamental group of its orbit space $M_n$ . For these n , $M_n$ is the complement of some knot in a closed 3-manifold $N_n$ , and an obvious 2-sheeted cyclic covering space of $M_n$ is homeomorphic to the $|2n|$-sheeted cyclic covering space of $S^3 - k$. We wonder whether the homology groups $H_1(M_n)$ would be interesting invariants of K. For 2-bridge knots in our standard setup we would take $A_0 = A\{g/2\}$, but for $\beta^2 \equiv +1 \pmod{2\alpha}$ we could also use $A\{(g+2)/4\}$. When $A_\infty$ contains glide-reflections we could play a similar game with J replacing $A_0$ in the above, and probably have fewer worries about torsion in $G_n$ too. In fact the two constructions can be combined when $A_\infty$ contains both glide-reflections and suitable EH-translations.

We conclude with some remarks about the minimal crossing number of a 2-bridge knot and the exceptional symmetry of the knots with $\beta^2 \equiv +1$. In [2] J.H. Conway defined a *rational knot* $1^*(a_1\ a_2\ \dots\ a_r)$ where the $a_j$ are nonzero integers. It has become folklore that an alternating rational knot $k^*$ corresponding to the 2-bridge normal form $k = (\alpha, \beta)$ is $1^*(a_1\ a_2\ \dots\ a_r)$ where

$$\frac{\alpha}{\beta} = a_1 + \frac{1}{a_2} + \frac{1}{a_3} + \dots + \frac{1}{a_r},$$

and that the fraction $\alpha/(\alpha - \beta)$ also works. J. Montesinos gave a beautiful discussion of this with pictures and proofs in [8]. This topic is of interest for several reasons and we shall have to refer to [8] for a proper account. Some of the material is due to Conway but he chose not to include it in [2]. The crossing number of $k^*$ is $a_1 + \dots + a_r$, and this number is widely expected to be the minimal crossing number $\nu$ for the type. There is also no suggestion that a proof or disproof of this for all 2-bridge knots is imminent. To be explicit, the number $\nu$ to be used in (4.2) is the crossing number of $k^*$, and if this standard conjecture turns out to be false we shall simply stop calling $\nu$ the minimal crossing number.

Given $(\alpha, \beta)$ let $\bar\beta$ be the integer such that $1 \le \bar\beta < 2\alpha$ and $\beta\bar\beta \equiv +1 \pmod{2\alpha}$. Then $(\alpha, \beta)$ and $(\alpha, \bar\beta)$ not only have the same type but they are also isotopic (no mirrors). If $k^* = 1^*(a_1\ a_2\ \dots\ a_r)$ is a rational model for $(\alpha, \beta)$ then $\bar k^* = 1^*(a_r, a_{r-1}, \dots, a_1)$ is a rational model for $(\alpha, \bar\beta)$. Thus there is an isotopy of $k^*$ on $\bar k^*$. The necessary and sufficient conditions for $k^* = \bar k^*$, i.e. $a_j = a_{r+1-j}$ $(j = 1, \dots, r)$, is $\beta = \bar{\beta}$ , which is the same as $\beta^2 \equiv +1 \pmod{2\alpha}$ . So for these knots the isotopy of $k^*$ on $k^{-*}$ becomes a symmetry of $k^*$ , and presumably this is the geometric interpretation of $A\{(g+2)/4\}$ .

## 5. Three bridge knots

Our calculations for the excellent hyperbolic structure for 3-bridge knots ran parallel, in principle, to those for 2-bridge knots, but there were awkward complications at every step. Some of these will make it impossible for even a strongly motivated reader to check all our assertions if his only electronic equipment is a programmable calculator. In addition, W. Thurston's announcement of a proof of a general existence theorem for hyperbolic structures reached us before we began analyzing the Ford domains $D(\pi K\theta)$ for the 3-bridge knots, and we took advantage of this by not completing the proof that the knots we examined are excellent. Hence the results below should be regarded as a description of the excellent hyperbolic structure based on the assumption that the structure actually exists.

The really hard part of the 3-bridge calculations is the determination of all the equivalence classes of p-reps of a 3-bridge knot group $\pi K$ . Finding all these classes boils down to setting up and solving a system of 4 nonlinear polynomial equations on 3 variables. There is yet no analogue to the palindromic symmetry $\varepsilon_i = \varepsilon_{\alpha-i}$ in the canonical presentation of a 2-bridge knot group that simplifies this task. Currently the only way to find all the solutions to this system (hence all the p-reps) is brute force, although we have found simplifications to the search for the excellent p-reps which are the only ones guaranteed to exist by Thurston's theorem. Only the brute force was available back in 1972 when the multivariate polynomial calculations were run on the Southampton computer using the SAC-1 file of Fortran programs that Professor G.E. Collins and his associates had developed over more than a decade. The solutions for the systems of equations were still incomplete when my Southampton lectureship expired, but they were brought to the current level of completeness at Strasbourg in 1973.

Our simplifications to the search for excellent p-reps amount to taking advantage of the observed symmetric of the knot, and the bigger the symmetry group the better. For example, the sequence of knots or links which are the closed braids $(\sigma_1 \sigma_2^{-1})^n$ for $n \geq 2$ admit an axis of rotational symmetry of order $n$ which reduces the search for the excellent p-reps to something that we did by hand quite easily. However, for the knots with $n \geq 4$ we expect that there is also an algebraic curve of p-reps (i.e. p-reps $\theta$ with Pdim $\theta = 1$ ), but this has not been verified even for the case $n = 4$ (the knot $8_{18}$ ), where the enormity of the calculations defeated the computer in 1972. If an excellent knot is invertible we can reduce the number of variables in the polynomial equations for the excellent p-reps to 2 . And if the knot has no symmetry we have no simplification.

There are also certain difficulties in the analysis of the Ford domain $D = D(\pi K \theta)$ for the excellent p-rep. We don't have a good rule to generate a list of elements $T_j \in \pi K \theta$ that will include all the side-pairing transformations of $D$ , and perhaps we didn't try hard enough to find one. What we actually did was a hybrid process that sometimes needed some persistence to complete, and an essential ingredient was an aggressive search for the transformations that are needed to complete the incomplete edge cycles of an early guess at $D$ . Another minor difficulty is that the closing trick does not always work, i.e. there may be an edge cycle whose cycle transformation contains more than 3 non EH-factors. But this only happened to us when the knots were very symmetric, and we found that the closing trick then worked smoothly for the subgroup $A^+$ of the normalizer $A$ of $\pi K \theta$ . Once we have proved that either one of $\pi K \theta$ , $A^+$ is discrete it follows automatically that the other is too, so it doesn't matter which group the closing trick succeeds for. The closing trick does work for the 4 examples below.

A final difficulty that we really have not overcome is the direct verification that the relations of the Poincaré presentation of $\pi K \theta$ are all images of relations in $\pi K$ itself, i.e. that $\theta$ is faithful. This seems unreasonably difficult to program for the computer, so it has been left to complete by hand. We did this for the simplest 3-bridge example, the knot $8_{20}$ (not reported on below), and we found this task so tedious that we never even attempted it for any other knot. The only part of our calculations of the excellent hyperbolic structure of most of our 3-bridge examples, including 3 of the 4 below, that is actually incomplete is just this step. We shall give the eager reader enough information for him to complete it by himself. This would not only avoid an appeal to Thurston's deep theorem, but it would also avoid requiring him to have faith in my assertion that further calculations not summarized here have excluded all other possibilities for an excellent hyperbolic structure. The additional trouble with the fourth knot, $9_{43}$ , is that we haven't really proved that the proposed excellent p-rep is actually a homomorphism. This is a consequence of numerical difficulties in the completion of the solution of the system of p-rep equations that defeated the Strasbourg computer. These almost certainly could have been overcome easily at Southampton after 1976, but we never got around to doing it.

The diagram of the knot k under consideration is drawn in part (a) of the Figure for the example, partly just to identify the knot unambiguously. But the main reason is to indicate the presentation for $\pi K$ that underlies the discussion. The meridian $x_j$ of $\pi k$ corresponds to a loop crossing under arc(j) by a left hand rule, and the meridian $\gamma \rightleftarrows x_1$ corresponds to a loop starting near arc(1) and running parallel, not antiparallel, to k. Some of our knot group presentations may be considerably worse than optimal, but we are not able to repeat the algebraic calculations of 1972-73 so we are stuck with them.

For each knot k the first step was to reduce the presentation of $\pi K$ from the diagram to a 3-bridge presentation of the shape

$$\pi K = |x_1, x_2, x_3 : w_1 x_1 = x_2 w_1, w_2 x_1 = x_3 w_2, w_3 x_2 = x_3 w_3|, \quad (5.1)$$

where $w_1, w_2, w_3$ are explicit words in $x_1, x_2, x_3$. I have sheets of computer printout dating perhaps to 1970 or 1971 listing the original presentations of $\pi K$ and the words $w_j$. The second step was to set up the system of p-rep polynomial equations, and we did this via the normal form

$$x_1 \mapsto A, \quad x_2 \mapsto B\{y_1\},$$

$$x_3 \mapsto C\{y_2, y_3\} := \begin{bmatrix} 1 - y_2 y_3 & y_2^2 y_3 \\ -y_3 & 1 + y_2 y_3 \end{bmatrix}. \quad (5.2)$$

There is no special merit in the normal form $C\{y_2 y_3\}$, and it has the minor disadvantage that the variable $y_2$ is not necessarily an algebraic integer for a p-rep $\theta$ in this form when $\theta$ is integral. But we can't do much better than (5.2) at present, and anyway we are stuck with it for our examples below. Remark that (5.2) will exclude any p-reps $\theta$ such that $x_1 \theta \rightleftarrows x_2 \theta$, but these can be found much more easily (by the commuting trick), and are certainly not faithful. It is a definite advantage to separate them out at the outset. To set up the p-rep equations we substituted the matrices (5.2) into the words $w_j$ of (5.1) to get matrices $W_j = w_j\{A, B, C\}$ in $\mathbb{Z}[y_1, y_2, y_3]$. We then substituted in the relations of (5.1) to get 3 matrix equations,

$$W_1A = BW_1 \ , \ W_2A = CW_2 \ , \ W_3B = CW_3 \ . \qquad (5.3)$$

Now any one of the 3 relations of (5.1) is a consequence of the other 2, and so the corresponding matrix equation can be omitted. We always omitted the one which we thought would give the most trouble. The same argument as before shows that each matrix equation unwinds to just 2 polynomial equations on the $y_i$ . Hence we get a system of 4 nonlinear polynomial equations on $y_1 , y_2 , y_3$ , the *p-rep equations* for (5.1).

Solving these was really grim, and our outline of our procedure cuts corners and omits complications. The basic idea was to eliminate $y_3$ from the p-rep equations, leaving a system of 3 equations in $\mathbb{Z}[y_1 , y_2]$ , and then eliminate $y_2$ from these to get 2 polynomial equations in $\mathbb{Z}[y_1]$ . Only Professor Collins' SAC system made this feasible, but it took some 8 months work to set the system up. Let $q(y_1)$ be the gcd of these last two polynomials, then we factored $q$ and for each factor $q_i$ we did the following. For each root $\xi_j^1$ of $q_i = 0$ we found a complex number $\xi_j^2$ such that $(\xi_j^1 , \xi_j^2)$ approximately solved the 3 polynomial equations on $y_1 , y_2$ , this step requiring only the ability to find all the roots of a polynomial in $\mathbb{C}[y_2]$ . We then found a triple $(\xi_j^1 , \xi_j^2 , \xi_j^3)$ which approximately solves the p-rep equations. Sometimes the required $\xi_j^2$ or $\xi_j^3$ did not exist, which meant that $q_i$ was an extraneous factor that does not correspond to p-reps of $\pi K$ . For a surviving factor $q_i$ we managed to determine each set of numbers $(\xi_j^\nu)$ , $\nu = 1 , 2 , 3$ , so accurately that we could guess an irreducible polynomial $f_\nu \in \mathbb{Z}[y]$ such that $f_\nu(\xi_j^\nu) = 0$ for all $j , \nu$ . Using $f_\nu$ we could use Newton's iteration to improve the accuracy of the $\xi_j^\nu$ to the machine limit. Guided by experience with the first examples we then solved for polynomials $g_1' , g_3' \in \mathbb{C}[y]$ such that $\xi_j^\nu = g_\nu'(\xi_j^2)$ , for $\nu = 1 , 3$ and all the $j$'s . We considered $g_1' , g_3'$ as approximations to polynomials $g_1 , g_3 \in \mathbb{Q}[y]$ , and by taking continued fraction expansions of the real parts of the coefficients of $g_1' , g_3'$ we tried to guess the coefficients of $g_1 , g_3$ . If successful we had a triple of polynomials $f_2 , g_1 , g_3$ in $\mathbb{Q}[y]$ which we could test to see if they really constituted a solution to the p-rep equations. To do this substitute $(g_1, y_2, g_3)$ for $(y_1, y_2, y_3)$ in each of the 4 original p-rep polynomial expressions, and determine whether the resulting polynomial in $\mathbb{Q}[y_2]$ is $\equiv 0 \pmod{f_2(y)}$. This always happened, and therefore $x_1 \mapsto A$, $x_2 \mapsto B\{\xi_j^1\}$, $x_3 \mapsto C\{\xi_j^2, \xi_j^3\}$ was proved to generate a p-rep $\theta = \theta\{\xi_j^2\}$ of $\pi K$. Unfortunately we didn't always manage to compute $g_1'$, $g_3'$ with sufficient accuracy to attempt a guess at $g_1$, $g_3$, and in these cases all we have are presumed p-reps $\theta\{\xi_j^1, \xi_j^2, \xi_j^3\}$ which we haven't proved to be homomorphisms. All the work on the p-rep equations following the application of SAC was done at Strasbourg.

5(a) The knot $8_{21}$

If we start from the labelled diagram Figure 4(a) of $k$, a model of $K = 8_{21}$, the setup we have just described leads to the following. The words $w_j$ of (5.1) are

$$w_1 = x_1 x_3^{-1} x_1^{-1} x_2^{-1} x_3^{-1} x_2 ,$$

$$w_2 = x_2 x_1 x_2^{-1} x_3 ,$$

$$w_3 = x_1^{-1} x_2^{-1} x_3^{-1} x_2 x_1^{-1} x_2^{-1} x_3 x_2 x_1 .$$

Using the $w_1$-, $w_2$-relations, (5.3) became equivalent to the assertion that the following polynomial expressions all vanish.

$$1 + y_1 + y_1^2 y_2 - y_3(1 + y_1 y_2)^2$$

$$2 - y_1 + y_1 y_2^2(-2 + y_1) + y_3(-1 - y_1 y_2 + y_1^2 y_2^3 + y_1^2 y_2^4)$$

$$-y_1^2 + y_3(y_1 + y_2(-1 + y_1 + y_1^2)) + y_2 y_3^2(1 + y_1 y_2)^2$$

$$1 + y_3(1+2y_2+y_1y_2^2) + y_3^2(-1-y_2(1+2y_1) - y_1y_2^2(3+y_1)-y_1y_3^2(1+2y_1)-y_1^2y_2^4) .$$

We then solved these equations by the outlined procedure, and the polynomial $q(y_1)$ obtained after elimination of $y_2$, $y_3$ and taking a gcd turned out to be

$$\{1 \ \ -1\}^2 . \{1 \ \ -3 \ \ 1\}^2 . \{4 \ \ 8 \ \ 1 \ \ -3 \ \ 1\} .$$

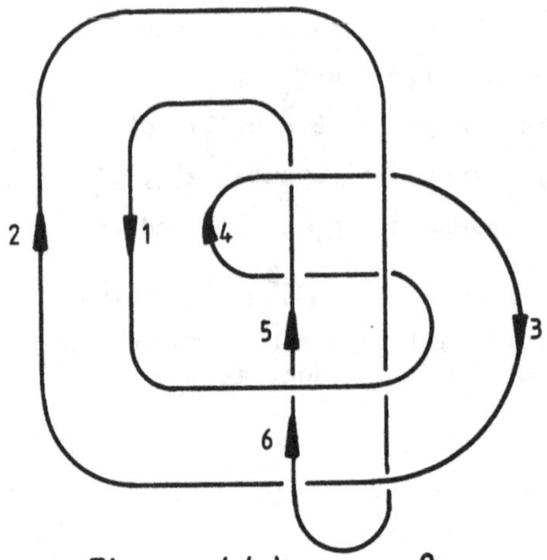


Figure 4(a). $8_{21}$

In $\pi K$, $X_4 = X_2^{-1} X_3 X_2$, $X_5 = X_4^{-1} X_1 X_4$,
$X_6 = X_1^{-1} X_5 X_1$,
$\gamma = X_2 X_5^{-1} X_2^{-1} X_6^{-1} X_1 X_3^{-1} X_1^{-1} X_4^{-1} X_1^{4}$.

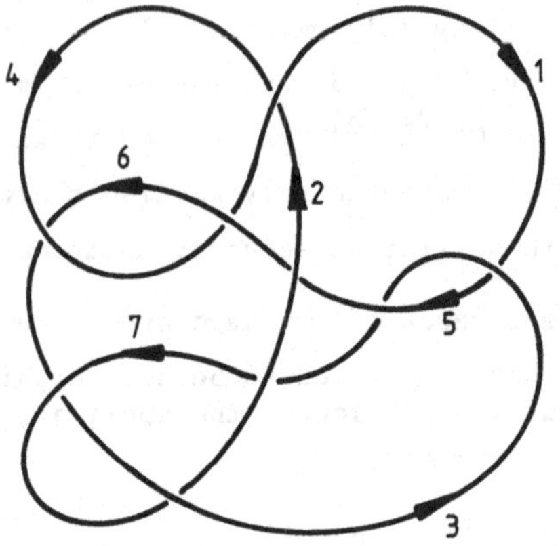


Figure 5(a). $9_{43}$

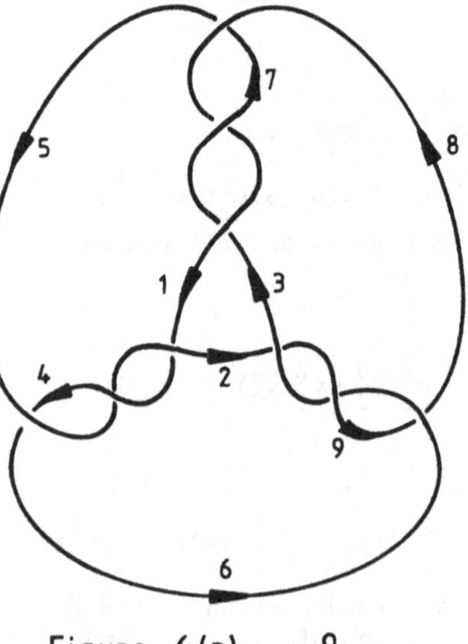


Figure 6(a). $9_{35}$

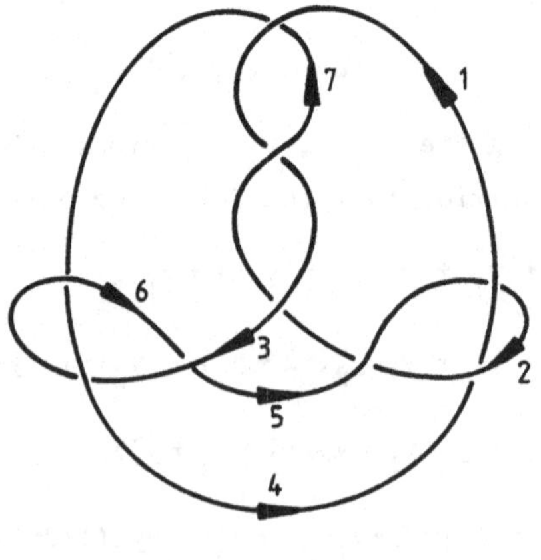


Figure 7(a). $9_{48}$

The linear factor proved to be extraneous, and the quadratic factor yielded p-reps $\theta$ such that $\pi K\theta$ contains the Hecke group $\Pi_5$ of [10]. There is also a p-rep on the modular group defined by $x_1\theta = x_2\theta = A$, $x_3\theta = B\{1\}$. Only the quartic factor could possibly correspond to an excellent p-rep, and the complete solution to the p-rep equations for it proved to be as follows. Let

$$f_2 := \{1 \quad 3 \quad 3 \quad 2 \quad 2\}\ ,$$
$$g_1 := \{6 \quad 7 \quad 4 \quad 6\}\ , \qquad g_3 := \{-7 \quad -7 \quad -4 \quad -6\} \tag{5.4}$$

Then these expressions constitute a solution, $(g_1(y_2), y_2, g_3(y_2))$, to the p-rep equations, and we write $\theta = \theta\{\xi\}$ for the p-rep of $\pi K$ corresponding to the root $\xi$ of $f_2(y_2) = 0$. We only noticed too late to redo the subsequent calculations that these p-reps themselves are not integral but they are simply equivalent to integral p-reps. It is not hard to show from the above information that the irreducible integral polynomial annihilating $a_{11}(x_3\theta)$ is

$$\{4 \quad 0 \quad -5 \quad 0 \quad 2\}\ ,$$

whence $a_{11}(x_3\theta)$ is not an algebraic integer. But the solution (5.4) shows that on conjugating by $A\{-y_2\}$ we get p-reps in the normal form

$$x_1 \mapsto A\ ,\ x_2 \mapsto \begin{bmatrix} 1+y_2^{-1} & 1 \\ -y_2^{-2} & 1-y_2^{-1} \end{bmatrix},\ x_3 \mapsto \begin{bmatrix} 1 & 0 \\ 1+y_2^{-2} & 1 \end{bmatrix}.$$

These are integral p-reps because the roots of $y^4 f_2(y^{-1}) = 0$ are algebraic integers. The ring for this normalized integral p-rep is $\mathbb{Z}[\xi^{-1}]$, and the longitude entry $g(\theta)$ is twice an element of $\mathbb{Z}[\xi^{-1}]$. The polynomial for $g/2$ is

$$\{92 \quad 76 \quad 35 \quad 8 \quad 1\}\ .$$

This is easier to write out than the longitude polynomial $Q(y) = 16Q_h(y/2)$ itself because its coefficients are much smaller. There is no functional equation for $Q(y)$ of the type of Theorem 1, so $S^3 - k$ admits no orientation-reversing autohomeomorphisms.

The approximate values for the excellent root $\xi$ of $f_2 = 0$ and of the longitude entry are

$$\xi \approx -0.58805 + 0.17228i \ , \ g \approx 3.43990 + 4.72374i \ .$$

Thus the computer used $A_1 = A\{g - 3\}$ as the second EH-side-pairing translation for $D = D(\pi K \theta)$ . It will turn out later that $R = R_0$ normalizes $\pi K \theta$ . The computer choose $D$ so that the non EH-side-pairing transformations are as follows.

| | cn(T) | cn($T^{-1}$) | rd(T) |
|---|---|---|---|
| $T_1 = x_4\theta$ | -0.412 - 0.172i | 0.368 - 1.211i | 0.531 |
| $T_3 = (x_1^{-1}\, x_2\, x_5^{-1})\theta$ | 0.478 - 0.692i | -0.478 + 0.692i | 0.729 |
| $T_5 = (x_2\, x_5\, x_2^{-1}\, x_1^{-1})\theta$ | $-\mathrm{cn}(T_1)$ | $-\mathrm{cn}(T_1^{-1})$ | $\mathrm{rd}(T_1)$ |
| $T_7 = T_5 \cdot (x_5^{-1}\, x_1)\theta$ | 0 | -0.280 + 2.362i | 0.613 |
| $T_9 = T_5 \cdot x_4^{-1}\theta$ | -0.198 - 1.670i | 0.198 + 1.670i | 0.729 |
| $T_{11} = A\, T_3\, T_7^{-1}\, A^{-1}$ | 0.308 + 2.190i | 0.088 + 1.151i | 0.531 |
| $T_{13} = T_7\, A^{-1}\, T_3^{-1}\, A$ | -0.033 + 0.432i | -0.247 + 1.930i | 0.515 |
| $T_{15} = R\, T_{11}\, R$ | $-\mathrm{cn}(T_{11})$ | $-\mathrm{cn}(T_{11}^{-1})$ | $\mathrm{rd}(T_{11})$ |
| $T_{17} = R\, T_{13}\, R$ | $-\mathrm{cn}(T_{13})$ | $-\mathrm{cn}(T_{13}^{-1})$ | $\mathrm{rd}(T_{13})$ |

The computer declared that the 83 non EH-edges of $D$ lie in 18 non EH-cycles whose cycle transformations are as follows.

| cycle | | cycle | |
|---|---|---|---|
| 1 | $A^{-1}\, T_3^{-1}\, A^{-1}\, T_3\, T_1$ | 2 | $A^{-1}\, T_5^{-1}\, T_9\, T_1$ |
| 3 | $T_7^{-1}\, T_9\, A^{-1}\, T_1$ | 4 | $A^{-1}\, T_7^{-1}\, A_1\, A^{-1}\, R_{15}^{-1}\, T_1$ |
| 5 | $A\, T_3^{-1}\, T_5\, A\, T_3$ | 6 | $T_7^{-1}\, A^{-1}\, T_{11}^{-1}\, A\, T_3$ |
| 7 | $A\, T_7^{-1}\, T_{13}\, T_3$ | 8 | $A\, T_9^{-1}\, T_{13}\, T_3$ |
| 9 | $T_{15}\, A_1^{-1}\, T_{11}^{-1}\, T_3$ | 10 | $A\, T_{15}\, A_1^{-1}\, T_7\, T_3$ |
| 11 | $T_{17}^{-1}\, T_9^{-1}\, A\, T_3$ | 12 | $A\, T_{17}^{-1}\, A_1^{-1}\, A\, T_7\, A\, T_3$ |
| 13 | $T_7^{-1}\, A_1\, A^{-1}\, T_9^{-1}\, A\, T_5$ | 14 | $A\, T_7^{-1}\, T_{11}^{-1}\, T_5$ |
| 15 | $T_{13}^{-1}\, A^{-1}\, T_9\, A_1^{-1}\, T_7$ | 16 | $T_{17}^{-1}\, A\, T_9^{-1}\, A\, T_7$ |
| 17 | $A\, T_9^{-1}\, T_{11}^{-1}\, T_9$ | 18 | $T_{15}^{-1}\, T_9^{-1}\, A^{-1}\, T_9$ |

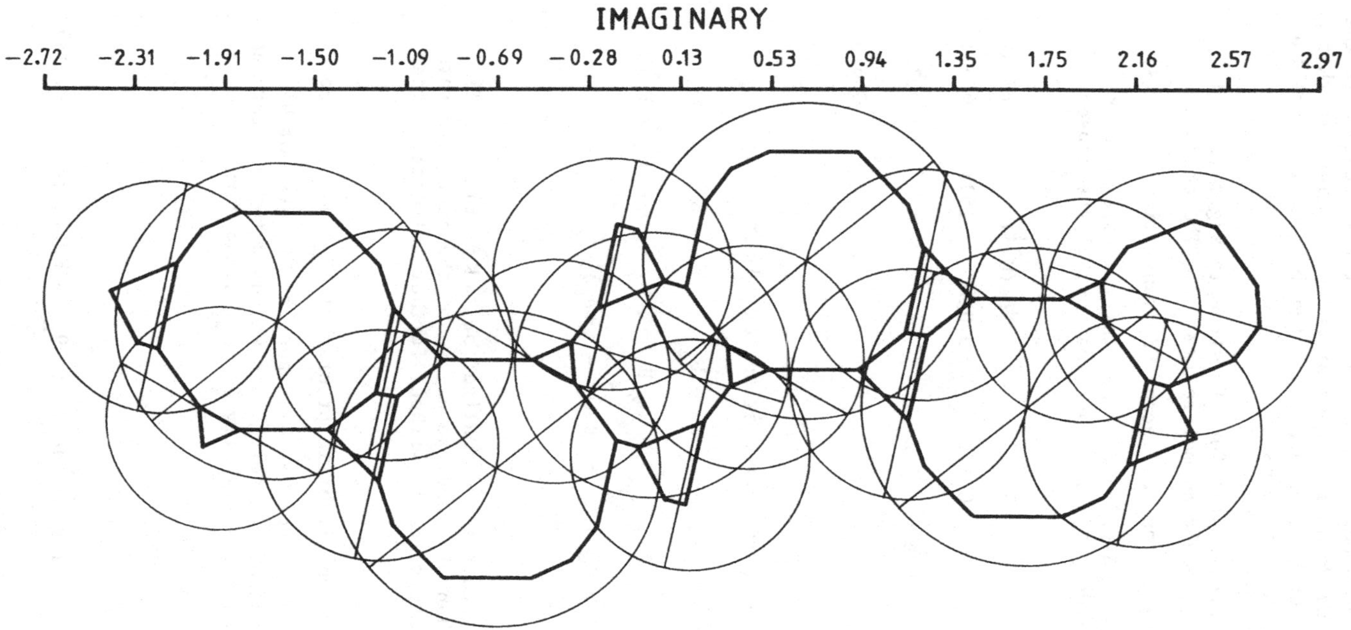
IMAGINARY
−2.72 −2.31 −1.91 −1.50 −1.09 −0.69 −0.28 0.13 0.53 0.94 1.35 1.75 2.16 2.57 2.97

Figure 4(b).

All the standard checks worked smoothly. All that we lack is the verification that the Poincaré presentation for D implies that $\theta$ is faithful, and we have provided all the information that the reader needs to do this for himself.

Inspection of Figure 4(b) for D strongly suggests that $R = R_0$ and $A\{g/2\}$ should normalize $\pi K \theta$. The computer choose $A_0 = A^{-1} A\{g/2\}$ as the second EH-translation for $D(A)$ where $A = \langle \pi K \theta , R , A\{g/2\} \rangle$. Then the following elements of $A$ are the non EH-side-pairing transformations of the chosen $D(A)$.

| | cn(U) | $\mathrm{cn}(U^{-1})$ | rd(U) |
|---|---|---|---|
| $U_1 = A_0^{-1} R T_1$ | -0.412 - 0.172i | -0.088 - 1.151i | 0.531 |
| $U_3 = R T_3$ | 0.478 - 0.692i | | 0.729 |
| $U_4 = A_0^{-1} T_7$ | 0 | | 0.613 |
| $U_5 = A_0^{-1} R T_{17}$ | 0.033 - 0.432i | | 0.515 |

The corresponding Poincaré presentation for $A$ is then:

$$R A U_1^{-1} U_3 U_1 = A^{-1} U_3 A U_3 R A_0 U_1 = U_4 U_3 A U_1 = E ,$$

$$R A U_4 A U_1^{-1} R A_0 U_1 = R A_0 A^{-1} U_3 U_5 U_3 = A U_4 U_5 A^{-1} U_3 = E ,$$

$$(U_4 R)^2 = U_3^2 = U_4^2 = U_5^2 = R^2 = (R A)^2 = (R A_0)^2 = E ,$$

$$A \rightleftarrows A_0 .$$

We can see directly that the normalizer of $\pi K \theta$ in $P\Gamma L(\mathbb{C})$ cannot be bigger than $A$, and the nature of the extra generators $R_0$, $A_0$ implies automatically without further calculation that $\mathrm{Out}(\pi K) \approx D_2$.

We are omitting an algebraic proof that $R_0$ and $A_0$ do indeed normalize $\pi K \theta$ because these assertions have not been challenged. One method for doing this was used twice before and another will be used for a symmetry of $9_{48}$, and either method could be used here.

5(b) The knot $9_{43}$

The principal novelty in the knot $9_{43} = K$ is that $\mathrm{Out}(\pi K)$

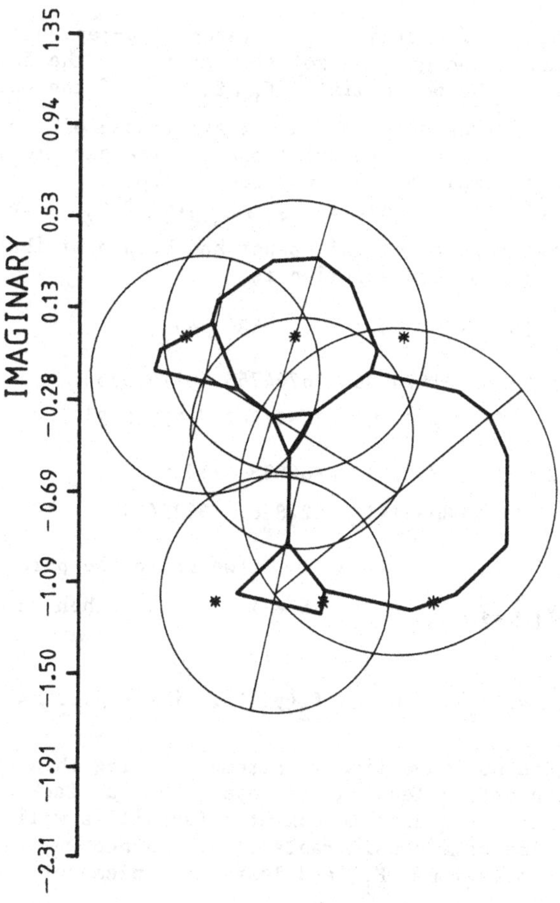


Figure 4(c).

is cyclic of order 2 , the only symmetry being a rotation of $S^3$ about an axis intersecting the knot. Starting from the presentation of $\pi K$ inferred from Figure 5(a) the words $w_2$ , $w_3$ of the corresponding presentation (5.1) are

$$w_2 = x_1^{-1} x_2 x_3 x_1 x_3^{-1} x_2 , \quad w_3 = x_1^{-1} x_3^{-1} x_2^{-1} x_3^{-1} .$$

We went about solving the corresponding system of p-rep equations in the standard manner and got the solution as far as the following incomplete version. The polynomials $f_1 , f_2 , f_3$ of the outlined method were found at Strasbourg, and a certain triple of their roots was used in the computer calculations for the excellent p-rep. But we only completed the subsequent step of the method, the determination of the polynomials $g_1 , g_3$ in $\mathbb{Q}[y_2]$ , in May 1980 after the first version of this paper had been distributed as a preprint. Our proposed solution is

$f_2$ = {1 -4 6 10 -30 9 49 -58 4} ,

$G_1$ = {1568352 -4516726 -1874206 16744756 -11525334 -24371039 34341342 -2391012} ,

$G_3$ = {1177513 -623881 -2843791 4805833 2107671 -8096131 5522886 -363548} ,

and $g_j = G_j/183067$ for $j = 1 , 3$ . If we write the p-rep equations as $F_i(y_1 , y_2 , y_3) = 0$ , $i = 1 , \ldots , 4$ , then we should have

$$F_i(g_1(y_2) , y_2 , g_3(y_2)) \equiv 0 (\mathrm{mod}\ f_2(y_2)) , \quad i = 1 , \ldots , 4 .$$

Unfortunately we did not have time to attempt proving this by direct substitution before leaving Princeton, and our lack of continued employment and access to computer facilities will delay our checking this for some considerable time. Hence we shall merely quote the expressions $F_i$ and leave this pleasant substitution to the more fortunate industrious readers.

$$F_1 = -y_1 - 2y_1y_2 - y_3 + y_1(y_2 + y_2^2)y_3 .$$

$$F_2 = -1 - y_2 + (2y_2 + y_1y_2^2 + y_1y_2^3)y_3 - y_1y_2^3y_3^2 .$$

$$F_3 = 1 + y_1 + y_1^2 + (1+2y_1-2(y_1+y_1^2)y_2)y_3 + (1-(1+2y_1)y_2+(y_1+y_1^2)y_2^2)y_3^2 .$$

$$F_4 = -y_1 + (1 - y_1+y_1y_2+y_1^2y_2^2)y_3 + (-2-y_1y_2^2-y_1^2y_2^3)y_3^2 + y_1y_2^2y_3^3 .$$

Note that we have not proved that our proposed p-reps are really homomorphisms, but we proceed on the assumption that they are (because K could not be excellent otherwise). Incidentally, the roots $\xi_j$ of $f_2(y_2) = 0$ are not algebraic integers, but the products $\xi_j g_3(\xi_j)$ and $(\xi_j)^2 g_3(\xi_j)$ are algebraic integers. Hence the p-reps $\theta\{\xi_j\}$ are integral. The longitude polynomial $Q(y)$ is $2^8 Q_h(y/2)/4$ where

$$Q_h = \{22429 \quad 404002 \quad 348023 \quad 180408 \quad 59659 \quad 12650 \quad 1665 \quad 124 \quad 4\} .$$

Because the top coefficient is 4 we see that the longitude parameter $g(\theta)$ is not quite twice a number in $R(\theta)$. By Theorem 2 $A\{g/2\}$ cannot possibly normalize $\pi K \theta$.

The triple $(g_1(\xi), \xi, g_3(\xi))$ for the excellent p-rep $\theta$ is approximately

$$(0.39003 + 4.94796i \;,\; 0.24976 - 0.28397i \;,\; 4.66409 + 2.21345i) ,$$

and the corresponding longitude parameter $g(\theta)$ is approximately $-9.72541 + 2.82977i$. The computer took $A_1 = A\{10 + g\}$ as the second EH-translation for the Ford domain $D(\pi K \theta)$. The non EH-side-pairing transformations for $D$ were chosen to be the following.

| | cn(T) | cn($T^{-1}$) | rd(T) |
|---|---|---|---|
| $T_1 = (x_1^{-1} x_6)\theta$ | 0.341 + 0.609i | 0 | 0.449 |
| $T_3 = (x_1^{-4} x_8)\theta$ | −0.085 + 0.928i | −0.140 + 1.333i | 0.523 |
| $T_5 = (x_1 x_3^{-1} x_2^{-1} x_1)\theta$ | −0.330 + 0.305i | −0.171 − 0.873i | 0.670 |
| $T_7$ | −0.093 − 0.257i | 0.340 + 1.395i | 0.408 |
| $T_9$ | 0.250 − 0.284i | 0.387 + 1.131i | 0.441 |
| $T_{11}$ | −0.407 − 0.311i | 0.434 + 0.866i | 0.408 |
| $T_{13}$ | 0.159 − 1.177i | 0.500 − 0.568i | 0.449 |

The missing expressions for the $T_j$ as words on the images of the meridians can easily be deduced from the Poincaré presentation of $\pi K \theta$. The computer found that $D$ has 60 non EH-edges which lie in 14 cycles that have the following cycle transformations.

| cycle | | cycle | |
|---|---|---|---|
| 1 | $T_3^{-1} A^{-1} T_9 T_1$ | 2 | $T_5^{-1} A_1^{-1} T_7 T_1$ |
| 3 | $A T_5^{-1} A^{-1} T_5 T_1$ | 4 | $T_{11} T_5 A^{-1} T_1$ |
| 5 | $T_3 T_5^{-1} A_1^{-1} T_3$ | 6 | $T_9 T_{13} A_1^{-1} A T_3$ |
| 7 | $A^{-1} T_9 T_7^{-1} T_3$ | 8 | $T_{11} T_7^{-1} A T_3$ |
| 9 | $A^{-1} T_{11} A^{-1} T_9^{-1} T_3$ | 10 | $A T_5^{-1} A^{-1} T_{13} T_5$ |
| 11 | $T_7^{-1} A_1 T_{13}^{-1} T_5$ | 12 | $A^{-1} T_9^{-1} T_7 T_5$ |
| 13 | $T_{11}^{-1} T_9 T_5$ | 14 | $A^{-1} T_{11} A^{-1} T_{13} A T_5$ |

If one were to check that the Poincaré presentation corresponding to this presents a group isomorphic to the knot group $\pi K$ we still wouldn't know that $\pi K \theta$ is discrete or that $\theta$ is a homomorphism. The trouble is that these relations are only known to hold approximately in $\pi K \theta$, and we need to know that they hold exactly when we are relying on the closing trick.

We turn to the normalizer $A \subset P\Gamma L(\mathbb{C})$ of $\pi K \theta$. The longitude polynomial $Q(y)$ satisfies no functional equation allowing glide-reflections to normalize $\pi K \theta$, so $A \subset PSL(\mathbb{C})$. From Figure 5(b) of D we see easily that $A_\infty$ contains no EH-translation not in $(\pi K \theta)_\infty$. This figure leads one to suppose that $R = R_t$ normalizes $\pi K \theta$, where

$$t = cn(T_3) + cn(T_3^{-1}) \approx -0.22588 + 2.26182i .$$

The computer found that this does work (to within the error tolerance) and choose D(A) so that its non EH-side-pairing transformations are as follows.

| | cn(U) | cn($U^{-1}$) | rd(U) |
|---|---|---|---|
| $U_1 = T_1$ | 0.341 + 0.609i | 0 | 0.449 |
| $U_3 = R\, T_3$ | -0.085 + 0.928i | | 0.523 |
| $U_4 = A_1^{-1} R\, T_5$ | -0.330 + 0.305i | | 0.670 |
| $U_5 = A\, R\, T_7$ | -0.093 - 0.257i | 0.434 + 0.866i | 0.408 |
| $U_7 = A\, R\, T_9$ | 0.250 - 0.284i | 0.387 + 1.131i | 0.441 |

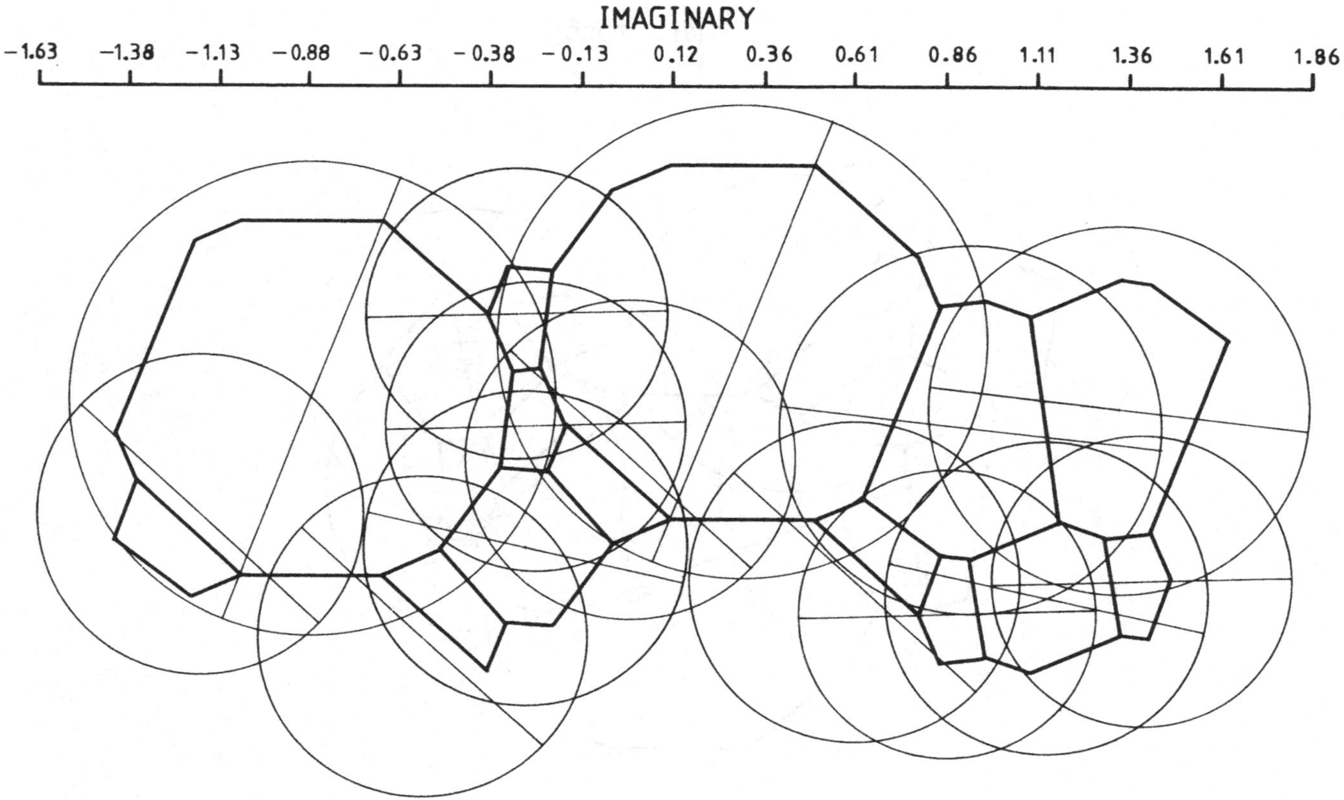
IMAGINARY
-1.63 -1.38 -1.13 -0.88 -0.63 -0.38 -0.13 0.12 0.36 0.61 0.86 1.11 1.36 1.61 1.86

Figure 5(b).

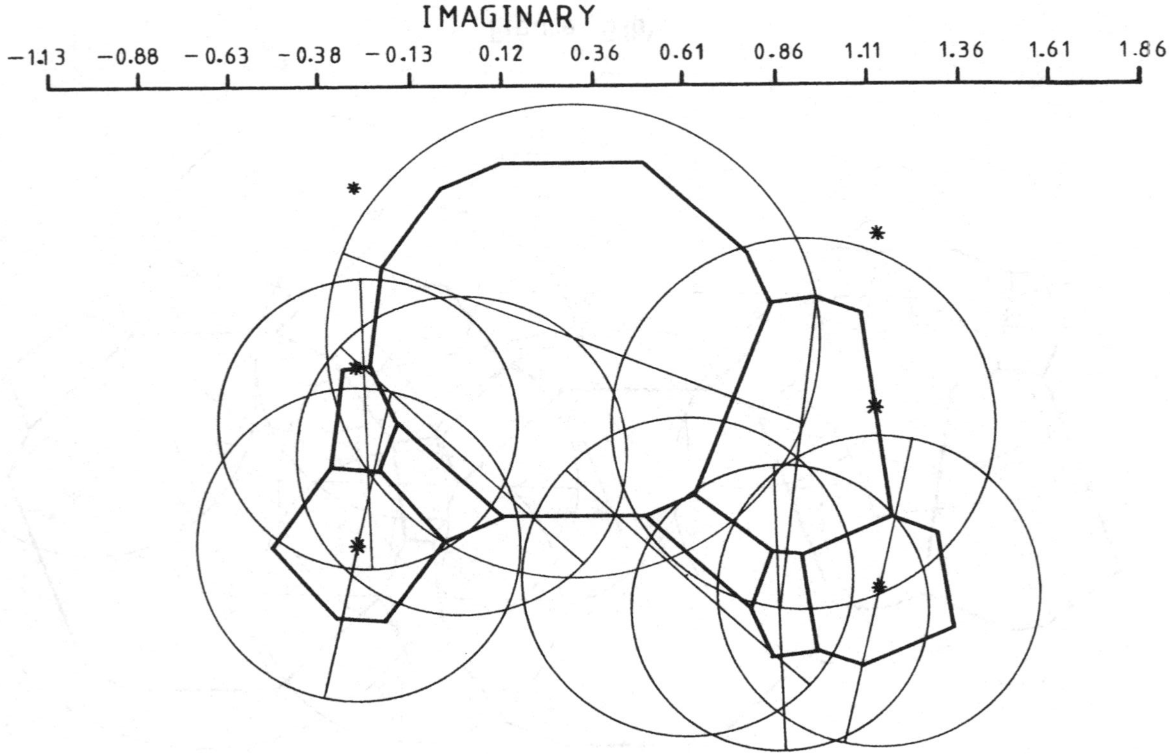


Figure 5(c).

The 32 non EH-edges fall into 11 cycles from which we get the following presentation for $A$ .

$$U_3U_7U_1 = U_4A^{-1}U_5U_1 = AU_4AU_4U_1 = U_5U_4A^{-1}U_1 = RU_3U_4U_3 = E ,$$

$$U_5RA_1U_5^{-1}U_3 = A^{-1}U_5U_7^{-1}RU_3 = RA_1U_5^{-1}U_7AU_4 = RA_1A^{-1}U_7^{-1}RA^{-1}U_7 = E ,$$

$$U_3^2 = U_4^2 = R^2 = (AR)^2 = (A_1R)^2 = E , \quad A \rightleftarrows A_1 .$$

This completes our analysis of $9_{43}$ .

5(c) The knot $9_{35}$

There are many common features of the p-rep sets of $9_{35}$ and $9_{48}$ , and we have an explanation for only one of them. This feature is a p-rep of the knot group on the subgroup of the Picard group $PSL(\mathbb{Z}[i])$ described by Figure 151 on p.432 of [3]. Contemplation of this led to the discovery of a general theorem about the p-reps of "triple skew unions" that I hope to write up for publication someday. There are also 3 completely mysterious classes of p-reps of each $\pi K$ whose abstract fields $F(\theta)$ are all equal to that for the p-reps of $6_1$ , the 2-bridge knot (9 , 5). Finally, the excellent class of p-reps for each knot has a cubic number field $F(\theta)$ , but the fields differ because the discriminants are 76 and 44 for $9_{35}$ and $9_{48}$ respectively. This implies that these knots are incommensurable. The Ford domains $D(\pi K \theta)$ have a common feature that should be pretty obvious from Figures 6(b) , 7(b) , viz. a symmetry of period 3. But there is a subtle difference between these symmetries which our analysis will make most apparent. We begin with $9_{35}$ because its symmetry allows a great collapse in the algebra for finding the excellent p-rep, and that of $9_{48}$ doesn't.

Our model $k$ of $K = 9_{35}$ has an obvious axis of rotational symmetry of order 3 , and we have chosen generators $x_1 , x_2 , x_3$ of $\pi K$ which are symmetric about this axis. If $\pi K$ admits an excellent p-rep $\theta$ then MRT implies that $\pi K \theta$ is normalized by an elliptic $S \subset PSL(\mathbb{C})$ such that $S^3 = E$ and

$$S\, x_1\theta\, S^{-1} = x_2\theta , \quad S^{-1}\, x_1\theta\, S = x_3\theta .$$

If we normalize $\theta$ to $x_1\theta = A$ , $x_2\theta = B\{y_1\}$ , then the first

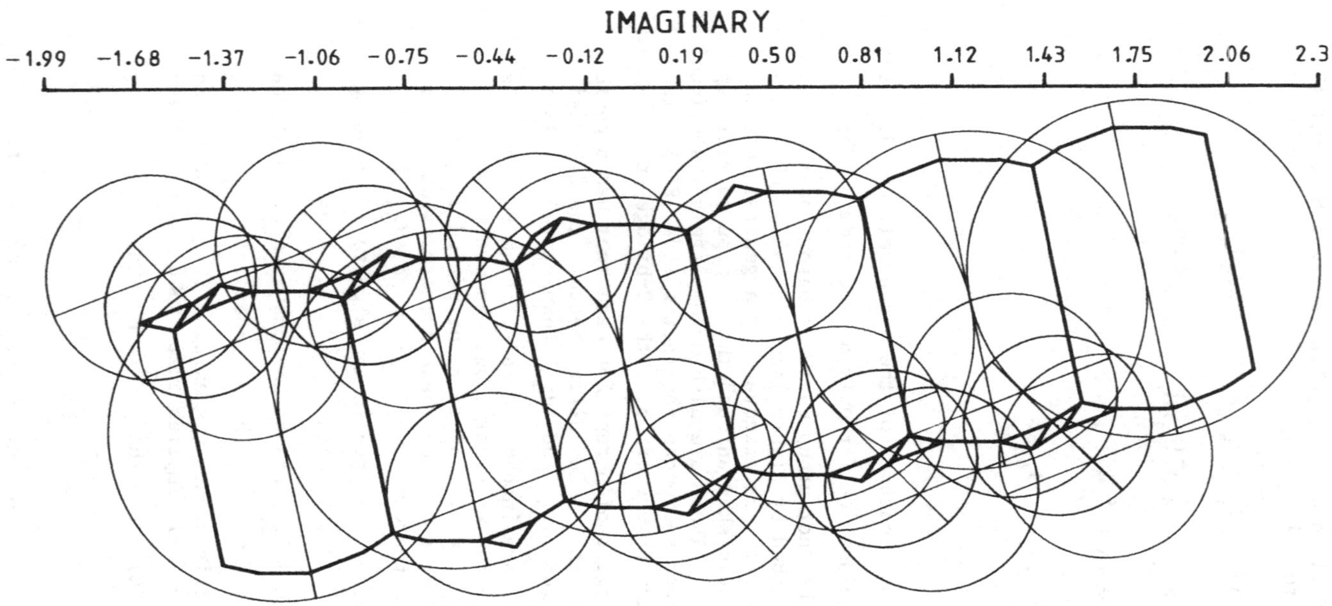
IMAGINARY
-1.99
-1.68
-1.37
-1.06
-0.75
-0.44
-0.12
0.19
0.50
0.81
1.12
1.43
1.75
2.06
2.37

Figure 6(b).

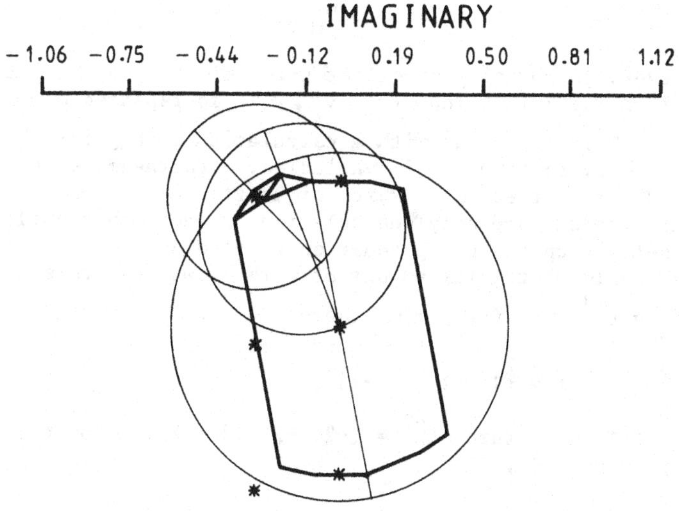


Figure 6(c).

two properties of S force

$$S = \begin{bmatrix} 0 & u^{-1} \\ -u & 1 \end{bmatrix}, \text{ where } u^2 = y_1 .$$

Then

$$S^{-1} A S = \begin{bmatrix} 1-u & 1 \\ -y_1 & 1+u \end{bmatrix},$$

and therefore every potentially excellent p-rep of $\pi K$ can be put in the normal form (5.2) in which $(y_1, y_2, y_3)$ is replaced by $(u^2, u^{-1}, u^2)$. This leaves us with a calculation with polynomials in $u$ alone, instead of 3 variables. Furthermore, the relations of $\pi K$ are sorted into orbits of length 3 by the rotational automorphism, and only one relation in each orbit needs to be used in setting up the p-rep equation for the excellent p-reps. Thus an almost trivial amount of work shows that this equation on $y_2 = u^{-1}$ is $f(y_2) = 0$ where

$$f = \{1 \quad -1 \quad 3 \quad -1\} .$$

Remark that if $f(\xi) = 0$ then $\xi^{-2} = \{-2 \quad -2 \quad 1\} (\xi)$. For the excellent p-rep $\theta = \theta\{\xi\}$,

$$\xi \approx 0.11535 - 0.58974i , \quad g(\theta) \approx -0.69212 + 3.53846i .$$

It turns out that $g(\theta) = -6\xi$, so the monic integral polynomial annihilating $g/6$ is

$$Q_6 = \{1 \quad 1 \quad 3 \quad 1\} .$$

Then the longitude polynomial $Q(y) = 6^3 Q_6(y/6)$ has odd degree, whence it does not admit a functional equation of the shape (1.2).

The computer found that the Ford domain $D(\pi K \theta)$ has 24 spherical sides, and it will be easy to check from the corresponding Poincaré presentation that the expressions of the corresponding side-pairing transformations as words on $x_1\theta, x_2\theta, x_3\theta$ are easily deduced from the edge cycle relations, provided that we know

$$T_1 = x_2\theta , \quad T_3 = x_3\theta , \quad T_{15} = (x_1^5 x_7^{-1} x_6^{-1} x_3^{-1} x_4^{-1} x_8^{-1} x_1^{-1})\theta .$$

| | cn(T) | $cn(T^{-1})$ | rd(T) |
|---|---|---|---|
| $T_1$ | -0.334 - 0.136i | 0.334 + 0.136i | 0.361 |
| $T_3$ | -0.219 - 0.726i | 0.450 - 0.454i | 0.361 |
| $T_5$ | 0 | -0.346 + 1.769i | 0.601 |
| $T_7$ | -0.319 - 1.633i | 0.319 + 1.633i | 0.361 |
| $T_9$ | -0.327 - 0.885i | 0.327 + 0.855i | 0.319 |
| $T_{11}$ | 0.104 + 1.316i | 0.435 + 1.043i | 0.361 |
| $T_{13}$ | -0.442 - 0.295i | 0.212 + 1.474i | 0.319 |
| $T_{15}$ | 0.115 - 0.590i | -0.231 + 1.179i | 0.601 |
| $T_{17}$ | 0.219 + 0.726i | -0.450 + 0.454i | 0.361 |
| $T_{19}$ | -0.104 - 1.316i | -0.435 - 1.043i | 0.361 |
| $T_{21}$ | 0.442 + 0.295i | -0.212 - 1.474i | 0.319 |
| $T_{23}$ | -0.115 + 0.590i | 0.231 - 1.179i | 0.601 |

Then the computer sorted the 88 non EH-edges of D into 20 cycles whose cycle transformations are as follows.

| cycle | | cycle | |
|---|---|---|---|
| 1 | $T_5^{-1} A^{-1} T_5 T_1$ | 2 | $A^{-1} T_5^{-1} A_1 T_{21} T_1$ |
| 3 | $T_{13}^{-1} A T_5 A^{-1} T_1$ | 4 | $T_9^{-1} A T_{15} A^{-1} T_3$ |
| 5 | $T_{15}^{-1} A^{-1} T_{15} T_3$ | 6 | $A^{-1} T_{15}^{-1} T_{13} A^{-1} T_3$ |
| 7 | $A T_5^{-1} A^{-1} T_7 A_1^{-1} T_5$ | 8 | $T_{13}^{-1} T_7 A_1^{-1} A T_5$ |
| 9 | $T_{15}^{-1} T_{23}^{-1} A_1^{-1} A T_5$ | 10 | $T_{21}^{-1} T_7^{-1} T_5$ |
| 11 | $T_{23}^{-1} T_{15}^{-1} T_5$ | 12 | $T_{15}^{-1} A^{-1} T_{11}^{-1} T_9$ |
| 13 | $T_{19} A^{-1} T_{23} T_9$ | 14 | $A^{-1} T_{23} A T_{17} T_9$ |
| 15 | $T_{13} A^{-1} T_{15}^{-1} T_{11}$ | 16 | $T_{15} A T_{15}^{-1} A^{-1} T_{11}$ |
| 17 | $T_{23}^{-1} A T_{23} T_{17}$ | 18 | $A T_{23}^{-1} T_{21} A T_{17}$ |
| 19 | $T_{21} A T_{23}^{-1} T_{19}$ | 20 | $T_{23} A^{-1} T_{23} A T_{19}$ |

Furthermore $A_1$ above is $AA\{g\}$. The direct verification that $\pi K\theta$ is isomorphic to $\pi K$ will be left to the reader, as usual.

Theorem 3 of §1 and the fact that the generators for $\pi K\theta$ are $S^{\nu} A S^{-\nu}$ for $\nu = -1, 0, 1$ imply that $9_{35}$ is invertible, and perhaps a similar argument shows $9_{48}$ is invertible. But in these cases we can see the inverting rotation of $S^3$ about an axis intersecting the knot directly from the pictures. For $9_{35}$ $R = R_0$ normalizes $\pi K\theta$, and there is a second rotation of order 2 of $S^3$ about an axis not intersecting $k$. This shows up in the normalizer $A$ of $\pi K\theta$ as $A\{g/2\}$, so

$$A\{g/6\} = A\{g/2\}\, A\{g/3\}^{-1}$$

normalizes $\pi K\theta$, as our calculations above led us to expect. Then $A = \langle \pi K\theta, R, A_0 \rangle$ where $A_0 = A\{g/6\}$ has the following non EH-generators.

| | cn(U) | rd(U) |
|---|---|---|
| $U_1 = R\,T_1$ | -0.334 - 0.136i | 0.361 |
| $U_2 = A_0^{-3}\,T_5$ | 0 | 0.601 |
| $U_3 = A^{-1}\,A_0\,T_{13}$ | -0.442 - 0.295i | 0.319 |

The Ford domain $D(A)$ has 10 non EH-edges which fall into 8 non EH-cycles, and the Poincaré presentation for $A$ is as follows.

$$U_2\,R\,A\,U_2\,U_1 = R\,A\,U_2\,U_3\,U_1 = E\,,$$

$$(A_0\,U_2)^3 = (U_2\,R)^2 = (U_3\,R\,A_0\,A)^2 = U_1^2 = U_2^2 = U_3^2 = R^2 = (RA)^2 = (RA_0)^2 = E\,,$$

$$A_0 \rightleftarrows A\,.$$

Also $\mathrm{Out}(\pi K) \approx D_6$ has order 12. This completes the discussion of $9_{35}$.

## 5(d) The knot $9_{48}$

Suppose one finds a free autohomeomorphism $\alpha$ of exact order $p \geq 3$ acting very nicely on $S^3$, such as a covering translation of the universal cover of the lens space $L(p, q)$. Let $D \subset S^3$ be an open fundamental domain for the action of $\alpha$, and select n points $P_1, \dots, P_n$ on bdry (D) . Locate their images $\alpha(P_j)$ , and choose n nonintersecting closed arcs which lie in D except for their endpoints, which are $\{P_j\}$ and $\{\alpha(P_j)\}$ . Consider the collection of all translates of these arcs by powers of $\alpha$ . These fit together to form a link $\ell \subset S^3$ which admits a free symmetry of order p that would seem quite weird to someone who wasn't shown how $\ell$ was constructed.

Bearing this in mind, we contemplate the model k of the knot type $K = 9_{48}$ shown in Figure 7(a). Any weird symmetry that k may possess does not, yet, permit the kind of simplification of the calculation of the excellent p-reps that makes the computer unnecessary. The words $w_j$ of (5.1) are

$$w_1 = x_3 x_1 x_2 x_1^{-1} x_2^{-1} x_3 x_2 , \quad w_2 = x_1 x_2^{-1} x_1^{-1} x_3^{-1} x_2^{-1} x_3 ,$$

and

$$w_3 = x_2 x_1 x_2^{-1} x_3^{-1} x_2 x_3 x_1 .$$

My records don't indicate which of the 3 relations was omitted. The potentially excellent polynomial solution $(g_1, y_2, g_3)$ (mod $f_2$) to the p-rep equations turned out as follows.

$$g_1 = \{-3 \;\; 4 \;\; -1\} , \quad g_3 = \{1 \;\; -6 \;\; 2\} ,$$

$$f_2 = \{-2 \;\; 4 \;\; -4 \;\; 1\} .$$

The corresponding p-reps $\theta\{\xi\} = \theta\{g_1(\xi) , \xi , g_3(\xi)\}$ are integral because the polynomials for $g_1(\xi)$ , $g_3(\xi)$ are

$$\{-1 \;\; 3 \;\; 1 \;\; 1\} , \quad \{-1 \;\; 11 \;\; 5 \;\; 1\}$$

respectively. The polynomial for $g/2$ , where $g = g(\theta)$ is the longitude parameter as usual, is

$$\{79 \;\; 33 \;\; 9 \;\; 1\} .$$

Hence $g/6$ and $g/3$ are not algebraic integers, and $A\{g/3\}$ cannot normalize $\pi K \theta$ . However, the annihilating polynomial for $(g+2)/6$ turns out to be

$$\{2 \;\; 2 \;\; 2 \;\; 1\} .$$

This suggests that $A\{(g+2)/6\}$ will normalize $\pi K \theta$ . The excellent triple $(g_1(\xi) , \xi , g_3(\xi))$ is, approximately,

$(-0.64780 + 1.72143i\ ,\ 0.58036 + 0.60629i\ ,\ -2.54369 - 2.23029i)$ ,

and

$$g(\theta) \approx -3.36893 + 6.69086i\ .$$

Hence the computer took $A_1 = A\{3+g\}$ as the second EH-transformation for the Ford domain $D = D(\pi K \theta)$ .

Then came the determination of $D$ itself. It is straightforward to check that the expressions of all of the non EH-side-pairing transformations of $D$ as words in $x_1\theta, \ldots, x_7\theta$ can be deduced from Figure 7(a) and the Poincaré presentation for $\pi K \theta$ provided that we know

$$A = x_1\,\theta\ ,\ T_1 = x_2\,\theta\ ,\ T_3 = (x_1^{-1} x_7\, x_1)\,\theta\ .$$

Then the data for the isometric spheres for $D$ is as follows.

| | cn(T) | $cn(T^{-1})$ | rd(T) |
|---|---|---|---|
| $T_1$ | -0.191 - 0.509i | 0.191 + 0.509i | 0.544 |
| $T_3$ | -0.420 + 0.606i | -0.037 + 1.624i | 0.544 |
| $T_5$ | 0.176 + 0.861i | 0.139 + 2.485i | 0.478 |
| $T_7$ | 0 | 0.316 + 3.345i | 0.737 |
| $T_9$ | -0.228 + 1.115i | 0.456 - 2.230i | 0.737 |
| $T_{11}$ | 0.124 + 2.837i | -0.124 - 2.837i | 0.544 |
| $T_{13}$ | 0.352 + 1.721i | -0.265 + 2.739i | 0.544 |
| $T_{15}$ | 0.265 - 2.739i | -0.352 - 1.721i | 0.544 |
| $T_{17}$ | 0.037 - 1.624i | 0.420 - 0.606i | 0.544 |
| $T_{19}$ | -0.280 - 3.091i | 0.052 - 1.976i | 0.478 |
| $T_{21}$ | -0.368 - 1.370i | -0.404 + 0.254i | 0.478 |
| $T_{23}$ | -0.456 + 2.230i | 0.228 - 1.115i | 0.737 |
| $T_{25}$ | -0.176 - 0.861i | -0.139 - 2.485i | 0.478 |
| $T_{27}$ | -0.052 + 1.976i | 0.280 - 3.091i | 0.478 |
| $T_{29}$ | 0.404 - 0.254i | 0.368 + 1.370i | 0.478 |

This data should be very helpful in interpreting our diagram, Figure 7(b), for $\mathcal{D}$. The polygons for the sphere of radius 0.478 are very nearly equilateral triangles. The computer sorted the 129 non EH-edges for $\mathcal{D}$ into 30 cycles and listed the following cycle transformations for them.

| cycle | | cycle | |
|---|---|---|---|
| 1 | $T_7^{-1} A^{-1} T_7 T_1$ | 2 | $A^{-1} T_{17} T_9 T_1$ |
| 3 | $T_{23} T_3 A^{-1} T_1$ | 4 | $A^{-1} T_{23} T_5 T_1$ |
| 5 | $T_{25}^{-1} T_9 A^{-1} T_1$ | 6 | $T_7^{-1} A T_{13} T_3$ |
| 7 | $A^{-1} T_7^{-1} A_1 A T_{27} T_3$ | 8 | $T_9^{-1} A^{-1} T_9 T_3$ |
| 9 | $T_{21} T_{23} A^{-1} T_3$ | 10 | $T_9^{-1} T_{23} A^{-1} T_5$ |
| 11 | $A T_9^{-1} T_{11} T_5$ | 12 | $A T_7^{-1} T_{11}^{-1} A_1^{-1} A^{-1} T_7$ |
| 13 | $T_9^{-1} T_{19} A^{-1} T_7$ | 14 | $T_{17} T_{15} A_1^{-1} T_7$ |
| 15 | $A^{-1} T_{17} T_{19} T_7$ | 16 | $T_{21} A^{-1} T_{23} A^{-1} T_7$ |
| 17 | $A T_{21} T_{15} A_1^{-1} A^{-1} T_7$ | 18 | $T_{23} T_{27}^{-1} A_1^{-1} T_7$ |
| 19 | $T_{29}^{-1} A T_9^{-1} A_1 T_7$ | 20 | $A^{-1} T_{29}^{-1} T_{13}^{-1} T_7$ |
| 21 | $A T_9^{-1} T_{15}^{-1} A^{-1} T_9$ | 22 | $T_{13}^{-1} T_{19}^{-1} A^{-1} T_9$ |
| 23 | $A^{-1} T_{13}^{-1} T_{11}^{-1} A^{-1} T_9$ | 24 | $T_{23}^{-1} T_{25}^{-1} A^{-1} T_9$ |
| 25 | $T_{29} T_{17} A^{-1} T_9$ | 26 | $T_{23}^{-1} A T_{25}^{-1} T_{11}$ |
| 27 | $A T_{23}^{-1} A T_{15} T_{11}$ | 28 | $A T_{23}^{-1} A^{-1} T_{23} T_{13}$ |
| 29 | $T_{27} A T_{23}^{-1} T_{15}$ | 30 | $T_{23} A^{-1} T_{23}^{-1} T_{17}$ |

The only missing step in our proof that $\theta$ is excellent is the proof that these relations are images of relations of $\pi K$ itself.

The hidden symmetry of $9_{48}$ corresponds to the automorphism of $\pi K \theta$ induced by conjugation by $A_3 := A\{(g+2)/3\}$, viz. $T \rightarrow A_3 T A_3^{-1}$. We need to prove that this conjugation sends $\pi K \theta$ on itself rather than onto some other group. We will do this indirectly by showing that $\pi K \theta$ admits an automorphism $\alpha$ whose effect on $\pi K \theta$ agrees with that of the conjugation. Then a supplementary argument using MRT will show that $\alpha$ is this

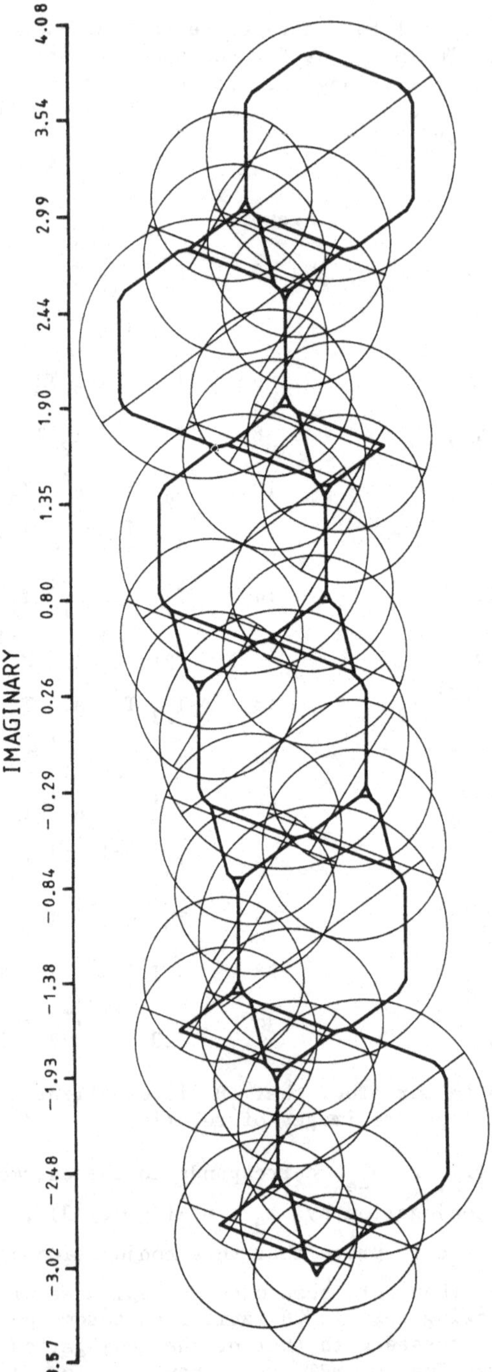
IMAGINARY
4.08
3.54
2.99
2.44
1.90
1.35
0.80
0.26
-0.29
-0.84
-1.38
-1.93
-2.48
-3.02
-3.57

Figure 7(b).

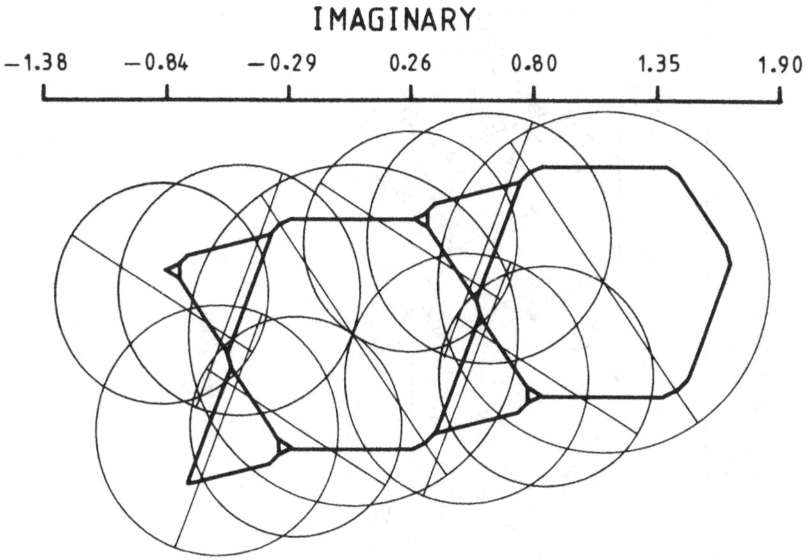


Figure 7(c).

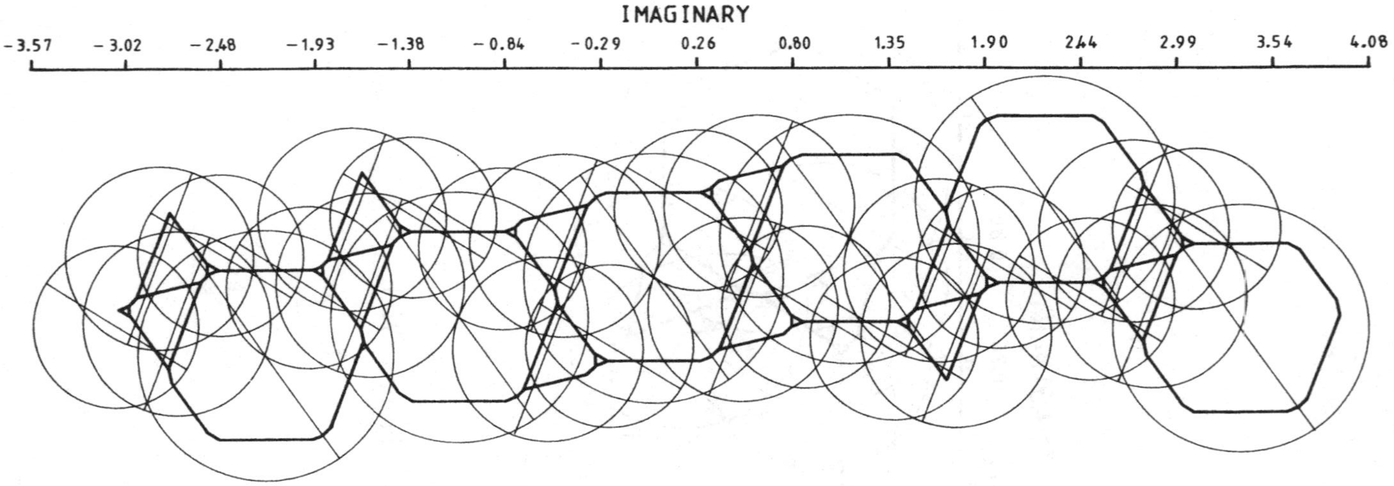
IMAGINARY
-3.57
- 3.02
-2.48
-1.93
-1.38
-0.84
-0.29
0.26
0.80
1.35
1.90
2.44
2.99
3.54
4.08

Figure 7(d).

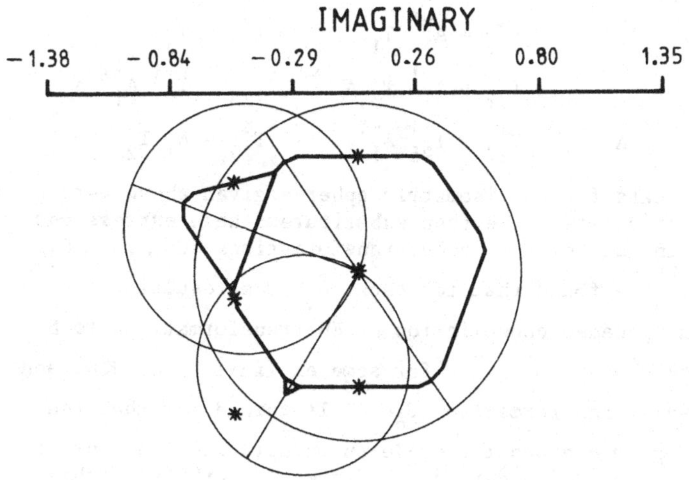


Figure 7(e).

conjugation by $A_3$. This will constitute a proof that k admits a weird symmetry which is not obviously described by diagrams. Writing $T^\alpha$ for $\alpha(T)$ where convenient, we define the effect of $\alpha$ on the generators by $A^\alpha = A$, $A\{g\}^\alpha = A\{g\}$, and the following list.

| | | |
|---|---|---|
| $T_1^\alpha = T_{13}\, A$ | $T_3^\alpha = A_1\, T_{11}\, A$ | $T_5^\alpha = A_1\, T_{19}$ |
| $T_7^\alpha = A_1\, T_{23}$ | $T_9^\alpha = T_7^{-1}\, A$ | $T_{11}^\alpha = A^{-1}\, T_{17}\, A_1^{-1}$ |
| $T_{13}^\alpha = A_1\, T_{15}\, A_1^{-1}$ | $T_{15}^\alpha = A^{-1}\, T_1$ | $T_{17}^\alpha = T_3$ |
| $T_{19}^\alpha = T_{21}\, A_1^{-1}$ | $T_{21}^\alpha = A^{-1}\, T_5\, A$ | $T_{23}^\alpha = T_9^{-1}\, A_1^{-1}\, A$ |
| $T_{25}^\alpha = A^{-1}\, T_{29}^{-1}\, A$ | $T_{27}^\alpha = T_{25}^{-1}\, A_1^{-1}$ | $T_{29}^\alpha = A_1\, T_{27}$ |

The numerical data for the isometric spheres given above were used in producing this list. We then substituted these expressions for the $T_j$ in each of the cycle transformations $C_1, \dots, C_{30}$ listed above. We found that for each $C_m$ the resulting expression only needed cancellation of EH-transformations to be reduced to the shape $VC_n^{\pm 1}V^{-1}$ for some explicit $V \in \pi K\theta$ and some other cycle transformation $C_n$. It turned out that the induced mapping $m \mapsto n$ on the cycle subscripts was a permutation of $\{1, 2, \dots, 30\}$ of order 3, and the 10 orbits of this permutation are as follows.

(1 28 21) (2 6 27) (3 23 14) (4 22 17) (5 20 29)

(7 26 25) (8 12 30) (9 11 15) (10 13 16) (18 24 19)

Hence the substitution $\alpha$ does define an automorphism of $\pi K\theta$.

According to MRT, there exists $U \in P\Gamma L(\mathbb{C})$ such that $\alpha$ agrees with the conjugation $T \longmapsto UTU^{-1}$ on $\pi K\theta$. Because $A^\alpha = A$, $A\{g\}^\alpha = A\{g\}$, $U$ must be $A\{h\}$ for some $h \in \mathbb{C}$. By tracing through the effect of the substitutions we find that $\alpha^3(T_1) = A\{2+g\}\, T_1\, A\{2+g\}^{-1}$, and this $= A\{3h\}\, T_1\, A\{3h\}^{-1}$. Hence $A\{\sigma\} := A\{2+g\}\, A\{3h\}^{-1}$ commutes with $T_1$. But if $\sigma \neq 0$ then $A\{\sigma\}$ is a parabolic fixing $\infty$ which cannot commute with a parabolic fixing 0 such as $T_1 = B\{f_1(\xi)\}$. Hence $A\{\sigma\} = E$, implying that $h = (2+g)/3$. Therefore $A_3$ normalizes $\pi K\theta$ and $S^3 - k$ admits a weird symmetry of order 3

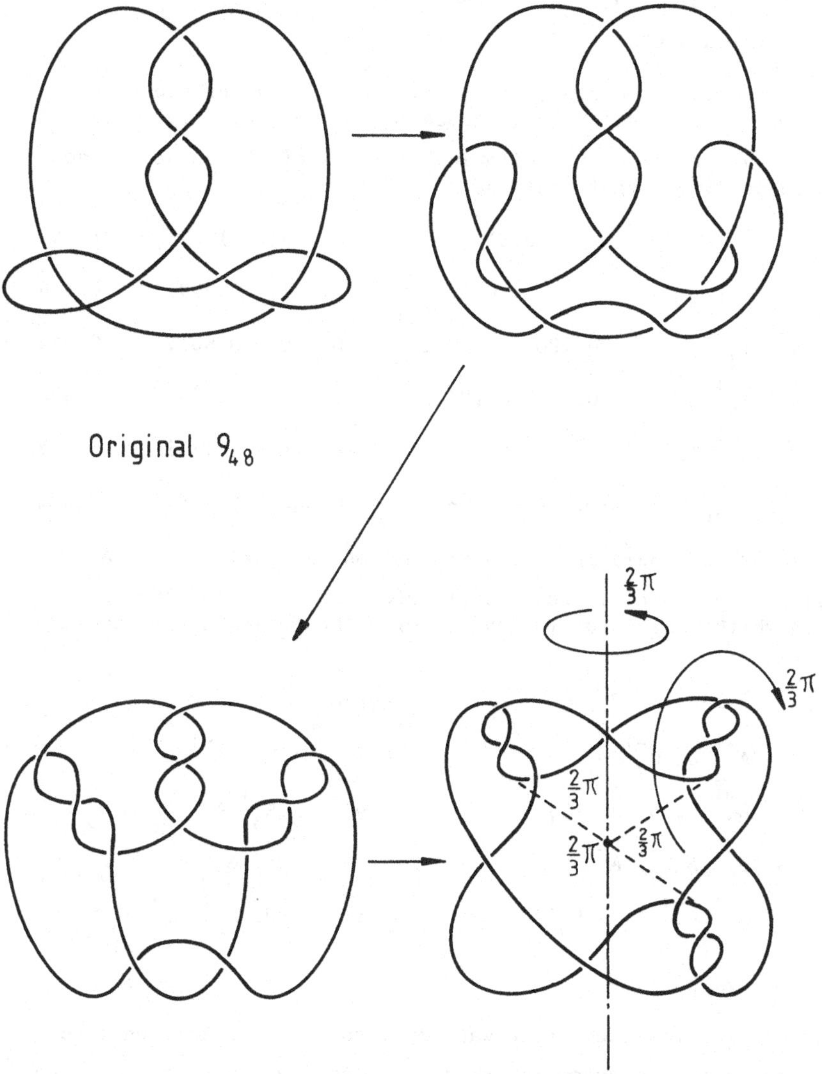


Figure 7(f). The exotic symmetry of $9_{48}$ represented as the product of two rotations of $S^3$ of angle $2\pi/3$ about different axes. These diagrams are copies of pictures drawn by Francis Bonahon.

that extends to an autohomeomorphism of $S^3$ . This symmetry can be realized as the product of two rotations of angle $\frac{2\pi}{3}$ about simply linked axes, cf. Figure 7(f).

We continue the study of $\alpha$ by considering the group $G = \langle \pi K \theta , A_3 \rangle$ and its orbit space $M(G)$ more closely. The computer volunteered the following elements of $G$ to act as non EH-side-pairing transformations of $D(G)$ .

| | cn(U) | $\mathrm{cn}(U^{-1})$ | rd(U) |
|---|---|---|---|
| $U_1 = T_1$ | -0.191 - 0.509i | 0.191 + 0.509i | 0.544 |
| $U_3 = A_3^{-1} T_3$ | -0.420 + 0.606i | 0.420 - 0.606i | 0.544 |
| $U_5 = A^{-1} A_3^{-1} T_5$ | 0.176 + 0.861i | -0.404 + 0.254i | 0.478 |
| $U_7 = A^{-1} A_3^{-1} T_7$ | 0 | -0.228 + 1.115i | 0.737 |
| $U_9 = A A_3 T_{25}$ | -0.176 - 0.861i | 0.404 - 0.254i | 0.478 |

The second EH-side-pairing transformation was taken to be $A_3$ itself. The computer found that $D(G)$ has 46 non EH-edge cycles which fall into 10 cycles with the following transformations.

| cycle | | cycle | |
|---|---|---|---|
| 1 | $A^{-1} U_3 U_7^{-1} U_1$ | 2 | $U_7^{-1} A^{-1} U_7 U_1$ |
| 3 | $A_3^{-1} U_7 U_3 A^{-1} U_1$ | 4 | $A_3^{-1} A^{-1} U_7 A U_5 U_1$ |
| 5 | $U_9^{-1} A U_7^{-1} A^{-1} U_1$ | 6 | $U_5 A U_7 A^{-1} U_3$ |
| 7 | $A^{-1} U_7^{-1} A_3 A U_9^{-1} U_3$ | 8 | $U_7 A^{-1} U_7^{-1} A_3 U_3$ |
| 9 | $U_7^2 U_5$ | 10 | $A_3^{-1} U_7 U_9 A_3^{-1} U_7$ |

The angle sum for each cycle was $2\pi$ , so $G$ is a torsion-free group and the autohomeomorphism of $S^3$ for $\alpha$ is a free mapping of period 3 . Hence $M(G) = N - k_1$ where $N$ is one of the lens spaces $L(3, 1)$ , $L(3, 2)$ , and $k_1$ is a knot in $N$ . Both lens spaces must occur, depending on whether we use $k$ or its mirror image.

We next consider the Alexander polynomial $D(t)$ of $k_1$ , or, more exactly, of $M(G)$ . It is easy to check that the abelianization of $G$ is isomorphic to $\mathbb{Z}$ , and that we may choose this isomorphism so that the images $a_\nu$ of our generators are as follows.

| | A | $U_1$ | $U_3$ | $U_5$ | $U_7$ | $U_9$ | $A_3$ |
|---|---|---|---|---|---|---|---|
| $a_\nu$ | 3 | 3 | 1 | -2 | 1 | 2 | 2 |

For the $\nu$-th generator in the above order we consider the matrix

$$X_\nu = \begin{bmatrix} t^{a_\nu} & b_\nu \\ 0 & 1 \end{bmatrix},$$

where the exponent of $t$ is $a_\nu$ , and $b_\nu \in \mathbb{C}$ . We then substitute these matrices $X_\nu$ for the generators of $G$ in our presentation to get a system of equations

$$S(t) \, . \, \vec{b} = \vec{0} \, .$$

Here $\vec{b}^{\dagger} = (b_1 , \ldots , b_7) \in \mathbb{C}^7$ and $S(t)$ is a $7 \times 11$ matrix over $\mathbb{Z}[t , t^{-1}]$ which presents the Alexander module of $M(G)$ . This presentation matrix can be boilded down to the $1 \times 2$ matrix

$$\|D(t) \; 0\| \, , \text{ where } D(t) = \{1 \;\; -1 \;\; -1 \;\; -1 \;\; 1\}$$

is the desired Alexander polynomial. The relation between $D(t)$ and the Alexander polynomial

$$\Delta(t) = \{1 \;\; -7 \;\; 11 \;\; -7 \;\; 1\}$$

of $9_{48}$ is

$$\Delta(t^3) = D(t) \, . \, D(\omega t) \, . \, D(\omega^2 t) \, , \text{ where } 1 + \omega + \omega^2 = 0 \, .$$

We conclude by describing the full automorphism group $A \subset P\Gamma L(\mathbb{C})$ of $\pi K \theta$ . First of all, $A \subset PSL(\mathbb{C})$ because the longitude polynomial $Q(y)$ has odd degree. The additional elements of $A_\infty$ needed to generate $A$ from $G$ are $R_0$ and $A_6 := A\{(g+2)/6\}$ . It is easy to see that $k$ has two axes of symmetry for $180^\circ$ rotations where one axis meets $k$ and the other doesn't, and therefore these new generators should be non controversial. The computer choose $D(A)$ so that its non EH-side-pairing transformations are as follows. Write $R$ for $R_0$ .

| | $cn(V) = cn(V^{-1})$ | $rd(V)$ |
|---|---|---|
| $V_1 = R\,U_1$ | $-0.191 - 0.509i$ | 0.544 |
| $V_2 = U_9\,R\,A_6$ | $0.404 - 0.254i$ | 0.478 |
| $V_3 = A_6^{-1}\,U_7$ | 0 | 0.737 |

The Poincaré presentation of $A$ deduced from the 15 non EH-edges of $D(A)$ is the following.

$$RA_6AV_1RA_6V_3A_6V_1 = RA_6V_2RA^{-1}V_3A_6AV_1 = V_3RAV_3V_1 = V_3RA_6^{-1}V_3V_2 = E ,$$

$$(V_3\,R)^2 = V_1^2 = V_2^2 = V_3^2 = R^2 = (RA)^2 = (RA_6)^2 = E ,$$

$$A \rightleftarrows A_6 .$$

The outer automorphism group for $\pi K \theta$ is again $D_6$, the same as for $9_{35}$. This completes our account of the symmetries of $9_{48}$.

## REFERENCES


1. G. BURDE, "Über periodische knoten", *Arch. Math.* 30 (1978), 487-492.

2. J. CONWAY, "An enumeration of knots and links, and some of their algebraic properties", Computational Problems in Abstract Alfebra, J. Leach Ed., *Pergamon Press*, (New York) 1970, 329-358.

3. R. FRICKE and F. KLEIN, *Vorlesungen Über die Theorie der Automorphen Functionen*, Band 1, B.G. Teubner (Leipzig) 1897.

4. R. HARTLEY and A. KAWAUCHI, "Polynomials of Amphicheiral Knots", *Math. Ann.* 243 (1979), 63-70.

5. T. JØRGENSEN, "Compact 3-manifolds of constant negative curvature fibering over the circle", *Ann. of Math.* 106 (1977), 61-72.

6. I. KRA, *Automorphic Forms and Kleinian Groups*, W.A. Benjamin (Mass.), 1972.

7. A. MARDEN, "The geometry of finitely generated Kleinian groups", *Ann. of Math.* 99 (1974), 383-462.

8. J.M. MONTESINOS, "Revêtements ramifiés de noeuds, espaces fibrés de Seifert et scindements de Heegaard", to appear in a future issue of *Asterisque* consecrated to dimensions 3 and 4.

9. R. RILEY, "Parabolic representations of knot groups. I", *Proc. London Math. Soc.* (3) 24 (1972), 217-242.

10. -------, "Hecke invariants of knot groups", *Glasgow Math. J.* 15 (1974), 17-26.

11. ------, "Parabolic representations of knot groups. II", *Proc. London Math. Soc.* (3) 31 (1975), 495-512.

12. ------, "A quadratic parabolic group", *Math. Proc. Cambridge Philos. Soc.* 77 (1975), 281-288.

13. -------, "Cubic parabolic groups", rejected for publication in 1975.

14. -------, "Discrete parabolic representations of link groups", *Mathematika* 22 (1975), 141-150.

15. -------, "An elliptical path from parabolic representations to hyperbolic structures", Topology of Low Dimensional Manifolds, R. Fenn Ed., *Springer Lecture Notes in Mathematics* 722 (1979).

16. -------, "Applications of a computer implementation of Poincaré's Theorem on Fundamental Polyhedra".

17. H. SCHUBERT, "Knoten mit zwei Brücken", *Math. Z.* 65 (1956), 133-170.

18. H. SEIFERT, "Komplexe mit Seitenzuordnung", *Göttingen Nachrichten* (1975), 49-80.

19. F. WALDHAUSEN, "On irreducible 3-manifolds which are sufficiently large", *Ann. of Math.* 87 (1968), 56-88.

# Part 3: Two-dimensional homotopy theory

# Identities among relations

R. BROWN and J. HUEBSCHMANN

## Introduction

An "identity among relations" is for a presentation of a group what a "syzygy among relations", as considered by Hilbert, is for a presentation of a module. The notion has ramifications in topology as well as in combinatorial group theory, and in particular is involved in some difficult problems in the algebraic topology of 2-diminsional complexes. Our aim is an exposition of this area explaining the connections with the following topics:

§1 Presentations and identities

§2 Pre-crossed and crossed modules

§3 Free crossed modules

§4 The associated chain complex

§5 Relationship with 2-dimensional CW-complexes

§6 Peiffer transformations

§7 Aspherical 2-complexes and aspherical presentations

§8 The identity property

§9 Examples and an unsettled problem of J.H.C. Whitehead

§10 Links and pictures

This will give also some background to the "two dimensional group theory" of the article [Br2] in this volume, and will contain all the prerequisites for the (posthumous) article by P. Stefan in this volume [St].

## 1. Presentations and identities

We consider a presentation $P = (X; R)$ of a group $G$. Thus we have a short exact sequence

$$1 \longrightarrow N \longrightarrow F \longrightarrow G \longrightarrow 1$$

where $F$ is the free group on the set $X$, $R$ is a subset of $F$ and $N = N(R)$ is the normal closure in $F$ of the set $R$. The group $F$ acts on $N$ by conjugation $c \mapsto c^u = u^{-1}cu$, for $c \in N$, $u \in F$, and the elements of $N$ are of course all *consequences* of the set $R$, that is any $c \in N$ is of the form

$$c = (r_1^{\varepsilon_1})^{u_1} (r_2^{\varepsilon_2})^{u_2} \ldots (r_n^{\varepsilon_n})^{u_n}$$

where $r_i \in R$, $\varepsilon_i = \pm 1$, $u_i \in F$.

An *identity among relations* is, heuristically, such a specified product in which $c = 1$ in $F$. A formal definition is given below, but let us first consider some examples.

EXAMPLE 1. For any elements $r, s$ of $R$ we have the identities

$$r^{-1}s^{-1}rs^r = 1$$

$$rs^{-1}r^{-1}s^{\tilde{r}} = 1 \qquad \text{where } \tilde{r} = r^{-1} .$$

These identities hold always, whatever $R$.

EXAMPLE 2. Suppose $r \in R$, $s \in F$ and $r = s^m$. Then $rs = sr$ (i.e. $s$ belongs to the centraliser $C(r)$ of $r$) and we have the identity

$$r^{-1}r^s = 1 .$$

However, for an element $r$ of a free group $F$ there is a unique element $z$ of $F$ such that $r = z^q$ with $q$ maximal, and then $C(r)$ is the infinite cyclic group on $z$. This element $z$ is called the *root* of $r$, and if $q > 1$, then $r$ is called a *proper power*. So if $r^{-1}r^s = 1$, then $s$ is a power $z^n$ of the root of $r$ (cf. [L-S], p.10, I.2.19).

EXAMPLE 3. Suppose the commutators $[x, y] = x^{-1}y^{-1}xy$, $[y, z]$, $[z, x]$ are among the relations. Then the well-known rule

$$[x, y][x, z]^y [y, z][y, x]^z [z, x][z, y]^x = 1$$

is an identity among the relations (since $[y, x] = [x, y]^{-1}$).

EXAMPLE 4. Consider the standard presentation $(x, y; r,s,t)$ of $\mathbb{Z}_2 \times \mathbb{Z}_2$ in which $r = x^2$, $s = y^2$, $t = x^{-1}y^{-1}xy$. We have an identity among relations

$$rts^{xy}(r^{-1})^y \, s^{-1}(t^{-1})^{\widetilde{yx}} = 1 ,$$

as is easily checked. This identity may be read as a path starting at 1 in the Cayley diagram:

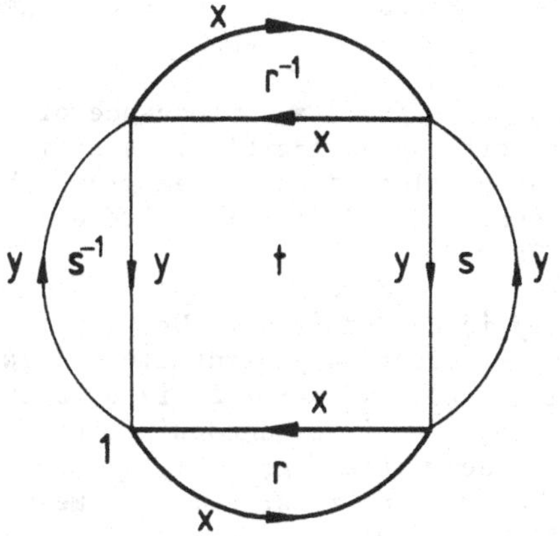


Figure 1

EXAMPLE 5. Consider the standard presentation $(x, y; r,s,t)$ of $S_3$, the symmetric group on three letters, in which $r = x^3$, $s = y^2$, $t = xyxy$. We have an identity among relations

$$ts^{-1}t(s^{-1})^{x}(r^{-1})^{\tilde{y}x}t^{x}(s^{-1})^{xx}r^{-1} = 1$$

as is easily checked. This identity may be read as a path starting at 1 in the Cayley diagram:

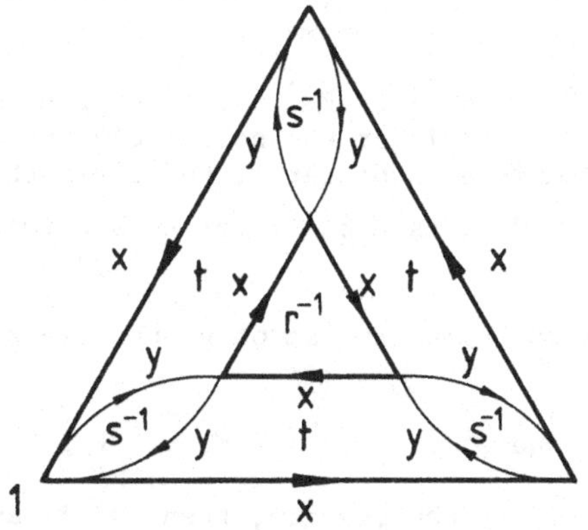


Figure 2

(Precise methods for obtaining the identities from these diagrams are given in §10).

Note that in these examples, conjugation is crucial. Indeed in the last example the elements $r$ , $s$ , $t$ freely generate a subgroup of $F$ .

The precise idea of *specifying* a consequence of the relations, and in particular of specifying an identity, is similar to that of specifying a relator as an element of a free group, but takes the action of $F$ into account. The definitions are due to Peiffer and Reidemeister [Pe, Re2].

One extra formality is needed first. We wish to allow for repeated relations, and so regard a presentation $P$ as a triple $(X; \mathsf{R}, w)$ where $\mathsf{R}$ is a set, $w: \mathsf{R} \longrightarrow F$ is a function to $F$ , the free group on $X$ , and $\mathsf{R}$ is assumed disjoint from $F$ . The elements of $\mathsf{R}$ will be written $\rho, \sigma, \tau, \ldots$ and the elements $w\rho, w\sigma, w\tau, \ldots$ will be written $r, s, t \ldots$ . We write $R = w(\mathsf{R})$ and $N = N(R)$ as above.

Let $H$ be the free operator group on $\mathsf{R}$ with right operators $(h, u) \longmapsto h^u$ , $h \in H$, $u \in F$, from $F$ . Thus as a group $H$ is free on the set $Y = \mathsf{R} \times F$ , the elements of which are written $\rho^u$, $\rho \in \mathsf{R}$ , $u \in F$ , with $\rho^1$ written $\rho$ and $(\rho^u)^{-1}$ written $\rho^{-u}$ or $(\rho^{-1})^u$ .

We mention an alternative way of obtaining $H$ (cf. [Pe; Satz 3 on p.69], [Me2]).

PROPOSITION 1. *Let* $F(X \cup \mathsf{R})$ *denote the group freely generated by* $X \cup \mathsf{R}$ . *Then* $H$ *is isomorphic to the normal closure of* $\mathsf{R}$ *in* $F(X \cup \mathsf{R})$ .

*Proof*. Clearly $F(X \cup \mathsf{R})$ contains $F$ , the free group on $X$ , and $F$ is a Schreier transversal for the normal closure of $\mathsf{R}$ in $F(X \cup \mathsf{R})$ . The Reidemeister-Schreier method gives the elements $u^{-1}\rho u$, $\rho \in \mathsf{R}$, $u \in F$ , as a basis for this normal closure. □

Let $\theta: H \longrightarrow F$ be the homomorphism of groups given on the basis elements by

$$\theta(\rho^u) = u^{-1}ru \ , \quad \text{where} \quad r = w\rho,\ \rho \in \mathsf{R},\ u \in F \ .$$

If we let $F$ act on itself by conjugation, then $\theta$ is an operator homomorphism, that is

$$\theta(h^u) = u^{-1}(\theta h)u,\ h \in H,\ u \in F \ .$$

Further $\theta(H) = N$ .

We now define the *identities among the relations* for the presentation $P = (X; R, w)$ to be the elements of the kernel $E$ of $\theta: H \longrightarrow F$.

However the group $E$ contains certain identities which are always present, namely those corresponding to the identities in Example 1. We therefore consider in $E$ the *basic Peiffer elements*, namely those of the form

$$p = a^{-1}b^{-1}ab^{\theta a}, \quad a, b \in Y .$$

More generally, any element of $H$ of the form

$$p = h^{-1}k^{-1}hk^{\theta h}, \quad h, k \in H$$

will be called a *Peiffer element*; such an element is an identity, i.e. belongs to $E$. These elements were introduced in [Pe; p.70] (not in the form given above) and [Re2].

The aim now is to factor out the Peiffer elements since these correspond to identities which are always present. It is convenient to discuss the situation in greater generality.

## 2. Pre-crossed and crossed modules

Let $\Gamma$ be a group. A *pre-crossed* $\Gamma$*-module* $(A, \delta)$ consists of a group $A$; a homomorphism $\delta: A \longrightarrow \Gamma$ of groups; and an action of $\Gamma$ on the right of $A$, written $(a, u) \longmapsto a^u$, $a \in A$, $u \in \Gamma$. A sole condition imposed is:

CM1) $\delta(a^u) = u^{-1}(\delta a)u$, $a \in A$, $u \in \Gamma$.

If we regard $\Gamma$ as acting on itself by conjugation, then CM1 says simply that $\delta$ is a $\Gamma$-morphism.

The pre-crossed $\Gamma$-module $(A, \delta)$ is a *crossed* $\Gamma$*-module* if it also satisfies

CM2) $a^{-1}ba = b^{\delta a}$, $a, b \in A$.

Let $(A, \delta)$, $(A', \delta')$ be pre-crossed $\Gamma$-modules. A *morphism of pre-crossed* $\Gamma$*-modules* $\phi: (A, \delta) \longrightarrow (A', \delta')$ is a $\Gamma$-morphism $\phi: A \longrightarrow A'$ of groups such that $\delta'\phi = \delta$.

We shall construct from any pre-crossed $\Gamma$-module a crossed $\Gamma$-module.

Let $(A, \delta)$ be a pre-crossed $\Gamma$-module. We call the elements

$$\langle a, b \rangle = a^{-1}b^{-1}ab^{\delta a}$$

for all $a, b \in A$ the *Peiffer elements* of $A$. We call the subgroup of $A$ generated by all Peiffer elements the *Peiffer group* of $(A, \delta)$. (We are here generalising a terminology used for the special case of the free pre-crossed $F$-module $(H, \theta)$ derived from a presentation. The Peiffer elements are then sometimes called *crossed commutators*; the elements of $P$ are sometimes called *Peiffer identities*; and the term *Peiffer group* is used in [Me2].)

PROPOSITION 2. *Let* $(A, \delta)$ *be a pre-crossed* $\Gamma$*-module. Then the Peiffer group* $P$ *of* $(A, \delta)$ *is normal in* $A$ *and* $\Gamma$*-invariant.*

*Proof.* Let $a, b, c \in A$. Then

$$c^{-1} < a, b > c = < ac, b > < c, b^{\delta a} >^{-1} .$$

Thus a conjugate of a Peiffer element is a product of Peiffer elements, and so $P$ is normal.

Let $a, b \in A$, $u \in \Gamma$. Then

$$< a, b >^{u} = < a^{u}, b^{u} >$$

(on using $\delta(a^{u}) = u^{-1}(\delta a)u$). So $P$ is $\Gamma$-invariant. □

COROLLARY. *Let* $(A, \delta)$ *be a pre-crossed* $\Gamma$*-module. Then there is a crossed* $\Gamma$*-module* $(C, \partial)$ *and a morphism* $\phi: (A, \delta) \longrightarrow (C, \partial)$ *of pre-crossed* $\Gamma$*-modules, such that* $\phi$ *is universal for morphisms from* $(A, \delta)$ *to crossed* $\Gamma$*-modules.*

*Proof.* Let $P$ be the group defined above. Then the quotient group $C = A/P$ is well-defined, and $C$ inherits a $\Gamma$-action and a $\Gamma$-morphism $\partial: C \longrightarrow \Gamma$. So $(C, \partial)$ is a pre-crossed $\Gamma$-module. By definition of $P$, we have $c^{-1}dc = d^{\partial c}$ for all $c, d \in C$, and so $(C, \partial)$ is a crossed $\Gamma$-module. The quotient morphism $A \longrightarrow C$ is clearly a morphism of pre-crossed $\Gamma$-modules and is universal for morphisms of $(A, \delta)$ to crossed $\Gamma$-modules. □

The above construction can be applied to the pre-crossed $F$-module $(H, \theta)$ constructed above from a presentation $\mathcal{P} = (X; R, w)$ of a group $G$. This will give a key example of a crossed $F$-module for $F$ a free group.

In this example, we shall be interested in the way the Peiffer group of $(H, \theta)$ is generated. The general result for this is the following.

PROPOSITION 3. *Let* $(A, \delta)$ *be a pre-crossed* $\Gamma$*-module and let* $V$ *be a set of generators for the group* $A$ *such that* $V$ *is* $\Gamma$*-invariant. Then the Peiffer group* $P$ *of* $(A, \delta)$ *is the normal closure in* $A$ *of the set* $Z$ *of Peiffer elements* $< a, b >$

*with* $a, b \in V$ .

*Proof.* Let $P'$ be the normal closure in $A$ of $Z$ . Then $P' \subset P \subset \text{Ker } \delta$ . The rule $< a, b >^u = < a^u, b^u >$ , $a, b \in A$ , $u \in \Gamma$ , shows that $Z$ , and hence also $P'$ , is $\Gamma$-invariant. So $C' = A/P'$ becomes a $\Gamma$-group and $\delta$ induces $\partial' : C' \longrightarrow \Gamma$ making $(C', \partial')$ a pre-crossed $\Gamma$-module.

Since $V$ generates $A$ as a group, we have

$$x^{\partial' y} = y^{-1}xy \qquad (*)$$

for all $x$ , $y$ in a set $V'$ which generates $C'$ as a group. For fixed $y$ , the set of $x$ satisfying $(*)$ is a subgroup of $C'$ , so $(*)$ is true for all $y \in V'$ , $x \in C'$ . Also the set of $y$ satisfying $(*)$ is closed under multiplication (because $x^{\partial'(yz)} = (x^{\partial' y})^{\partial' z} = z^{-1}(x^{\partial' y}) z = z^{-1}y^{-1}xyz$) and under inversion (because if $x^{\partial' y^{-1}} = w$ , then $x = w^{\partial' y} = y^{-1}wy$ , so $w = yxy^{-1}$). It follows that $(*)$ holds for all $x$ , $y \in C'$ , and hence $P \subset P'$ . □

COROLLARY. *Let* $(H, \theta)$ *be the pre-crossed F-module derived from a presentation* $\mathcal{P} = (X; R, w)$ . *Then the Peiffer group* $P$ *of* $(H, \theta)$ *is the normal closure in* $H$ *of the basic Peiffer elements* $< a, b >$ , $a$ , $b \in R \times F$ . □

We have now constructed a useful family of crossed F-modules, namely, those derived from a presentation. These examples will be fundamental in later pages. Other examples of crossed $\Gamma$-modules are:

(i) $(A, i)$ , in which $i$ is the inclusion of a normal subgroup $A$ of $\Gamma$ and $\Gamma$ acts on $A$ by conjugation,

(ii) $(A, 0)$ in which $A$ is a $\Gamma$-module in the usual sense and $0$ is the constant map,

(iii) $(A, \delta)$ , where $A$ is a group, $\Gamma = \text{Aut } A$ acts on $A$ in the obvious way, and $\delta : A \longrightarrow \text{Aut } A$ assigns to $a$ in $A$ the inner automorphism $x \longmapsto a^{-1}xa$ of $A$ .

Thus a crossed module generalises the concepts of a normal subgroup and that of an ordinary module.

We shall need some basic algebraic properties of crossed modules.

Let $(A, \partial)$ be a crossed $\Gamma$-module. We write $\pi$ for $\text{Ker } \partial$ and $N$ for $\text{Im } \partial$ . So we have an induced exact sequence of $\Gamma$-groups

$$1 \longrightarrow \pi \longrightarrow A \xrightarrow{\partial'} N \longrightarrow 1 \qquad (*).$$

We can now state some easy properties of these groups.

(2.1) $N$ *is normal in* $\Gamma$ *so that we can set* $G = \operatorname{Coker} \partial$ *to obtain an exact sequence of groups*

$$1 \longrightarrow N \longrightarrow \Gamma \longrightarrow G \longrightarrow 1 .$$

(2.2) $\pi$ *is contained in the centre* $ZA$ *of* $A$, *and in particular* $\pi$ *is abelian*.

(2.3) *The subgroup* $N$ *of* $\Gamma$ *acts trivially on* $ZA$, *and so also on* $\pi$ ; *hence* $\pi$ *inherits an action of* $G = \Gamma/N$ *to become a* G-*module*.

(2.4) *The abelianised group* $\bar{A} = A/[A, A]$ *inherits a structure of* G-*module*.

This last result is proved by noting that $N = \partial A$ acts trivially on $\bar{A}$ since for $a, b \in A$ the element $(b^{-1})^{\partial a} b$ is the commutator $a^{-1}b^{-1}ab$, by (CM2).

Since $N$ is normal in $\Gamma$, the action of $\Gamma$ on $N$ by conjugation determines an action of $G$ on the abelianised group $\bar{N} = N/[N, N]$, so that $\bar{N}$ becomes a G-module. It is clear that (*) determines an exact sequence of G-modules

$$\pi \longrightarrow \bar{A} \longrightarrow \bar{N} \longrightarrow 0 .$$

In general the map $\pi \longrightarrow \bar{A}$ is not injective. To see this, consider a group $A$ and the crossed (Aut A)-module $(A, \delta)$ of (iii) above. Then $\pi = \operatorname{Ker} \partial$ is the centre $ZA$ of $A$. There are non-abelian groups $A$ such that $1 \neq ZA \subset [A, A]$, for example the quaternion group, the dihedral groups $D_{2n}$, and many others. For all these, the composite $\pi = ZA \longrightarrow A \longrightarrow \bar{A}$ is trivial, and so not injective. This example gives point to the following result, which uses the notation of previous paragraphs.

PROPOSITION 4. *If the exact sequence* (*) *has a section (a group homomorphism but not necessarily a* $\Gamma$-*map), then* $A$ *is isomorphic as group to* $\pi \times \bar{N}$. *Further* $[A, A] \cap \pi = \{1\}$, *the induced map* $\pi \longrightarrow \bar{A}$ *is injective, and so the sequence*

$$0 \longrightarrow \pi \longrightarrow \bar{A} \longrightarrow \bar{N} \longrightarrow 0$$

*is a short exact sequence of* $\Gamma$-*modules*.

*Proof*. Let $s\colon N \longrightarrow A$ be a section of (*). Then $A = \pi \times sN$ (since $\pi$ is in the centre of $A$). Since $\pi$ is abelian, $[A, A] = [sN, sN]$ whence $[A, A] \cap \pi = \{1\}$. This implies $\pi \longrightarrow \bar{A}$ injective. □

These results apply to the crossed F-module $(C, \partial)$ derived as in §1 from a presentation $P = (X; R, w)$. In this case $F$ is a free group and hence so also is $N = \partial C$. The sequence $1 \longrightarrow \pi \longrightarrow C \xrightarrow{\partial'} N \longrightarrow 1$ has a section, and so Proposition 4 applies. The G-module $\bar{N}$ has been much studied - it is called the *relation module* of $P$ (see [D1-4, G2, L1-2, We]). We will see more of $\bar{N}$ later.

In the next section we present an important universal property of the crossed F-module $(C, \partial)$ of a presentation $P$.

## 3. Free Crossed modules

Given a presentation $P = (X; R, w)$ we constructed in §1 a pre-crossed F-module $(H, \theta)$, where $F$ is the free group on $X$. From this we can construct a crossed F-module $(C, \partial)$. It is convenient to give these constructions in greater generality.

Let $(A, \delta)$ be a pre-crossed $\Gamma$-module, let $R$ be a set and let $v: R \longrightarrow A$ be a function. We say $(A, \delta)$ is a *free pre-crossed* $\Gamma$*-module with basis* $v$ if for any pre-crossed $\Gamma$-module $(A', \delta')$ and function $v': R \longrightarrow A'$ such that $\delta'v' = \delta v$, there is a unique morphism $\phi: (A, \delta) \longrightarrow (A', \delta')$ of pre-crossed $\Gamma$-modules such that $\phi v = v'$. In such case, we also emphasise the rôle of the function $w = \delta v: R \longrightarrow \Gamma$ by calling $(A, \delta)$, with the function $v$, a *free pre-crossed* $\Gamma$*-module on* $w$. If $(A, \delta)$ is a crossed $\Gamma$-module, and $(A, \delta)$ with $v$ has the above universal property for maps into crossed $\Gamma$-modules, then we call $(A, \delta)$ a *free crossed* $\Gamma$*-module with basis* $v$ (or *on* $w$).

PROPOSITION 5. *Let* $\Gamma$ *be a group,* $R$ *a set and* $w: R \longrightarrow \Gamma$ *a function. Then a free pre-crossed* $\Gamma$*-module on* $w$, *and a free crossed* $\Gamma$*-module on* $w$, *exist, and are each uniquely determined up to isomorphism.*

*Proof.* For the existence of a free pre-crossed $\Gamma$-module we generalise easily the construction of §1. That is, we let $H$ be the free group on the set $R \times \Gamma$, and write the elements of this set as $\rho^u$, $\rho \in R$, $u \in \Gamma$. Let $\Gamma$ act on $H$ by acting on the generators as $(\rho^u)^v = \rho^{uv}$, $\rho \in R$, $u, v \in \Gamma$. Define $\theta: H \to \Gamma$ by its values on the generators

$$\theta(\rho^u) = u^{-1}(w\rho)u , \quad \rho \in R , \ u \in \Gamma .$$

Define $v: R \longrightarrow H$ by $v(\rho) = \rho^1$, $\rho \in R$, so that $\delta v = w$. Then $(H, \theta)$ is a pre-crossed $\Gamma$-module, which, with $v$, is easily checked to be a free pre-crossed $\Gamma$-module on $w$.

From $(H, \theta)$ we can form a crossed $\Gamma$-module $(C, \partial)$ by factoring out the Peiffer elements. Then $(C, \partial)$ with the composite $R \longrightarrow H \longrightarrow C$ is a free crossed $\Gamma$-module on $w$.

The uniqueness of these constructions up to isomorphism follows by the usual universal argument. □

The definition and construction of a free crossed $\Gamma$-module is due to Whitehead [Wh3]. We have followed a suggestion of P.J. Higgins and shown the rôle of the pre-crossed modules, since they are used implicitly also in some later proofs.

It is clear from the construction of the free pre-crossed $\Gamma$-module $(H, \theta)$ on $w : R \longrightarrow \Gamma$ that the basis function $v : R \longrightarrow H$ is injective. We wish to have this result for the free crossed $\Gamma$-module.

PROPOSITION 6. *Let* $(C, \partial)$ *be the free crossed* $\Gamma$*-module on* $w : R \longrightarrow \Gamma$, *with basis function* $v : R \longrightarrow \Gamma$. *Then* $v$ *is injective.*

This is most easily proved using the following result.

PROPOSITION 7. *If* $(C, \partial)$ *is a free crossed* $\Gamma$*-module, with basis* $v : R \longrightarrow C$, *and* $G = \text{Coker}\ \partial$, *then the abelianised group* $\overline{C}$ *is a G-module that is free on the composition* $\overline{v} : R \xrightarrow{v} C \longrightarrow \overline{C}$. *Hence* $\overline{v}$, *and so also* $v$, *is injective.*

*Proof.* By (2.4), $\overline{C}$ has the structure of G-module.

Let $p : \Gamma \longrightarrow G$ be the quotient map, and let $M$ be a G-module. Then $\Gamma \times M$, with projection to $\Gamma$, becomes a crossed $\Gamma$-module with action of $\Gamma$ by conjugation on $\Gamma$ and *via* $p$ on $M$.

Let $v' : R \longrightarrow M$ be a function. Define $v'' = (\partial v, v') : R \longrightarrow \Gamma \times M$. Freeness of $C$ gives a morphism $\phi : C \longrightarrow \Gamma \times M$ of crossed $\Gamma$-modules such that $\phi v = v''$. Composition with projection gives $\phi' : C \longrightarrow M$, a morphism of groups which factors through $\overline{\phi} : \overline{C} \longrightarrow M$. This is a G-morphism as required. □

We can now regard $v$ and $\overline{v}$ as inclusions, and link Proposition 7 nicely with Proposition 4.

COROLLARY. *Let* $(C, \partial)$ *be the free crossed* F*-module constructed as above from a presentation* $(X; R, w)$ *of a group* $G$. *Write* $\pi = \text{Ker}\ \partial$, $N = \text{Im}\ \partial$. *Then the induced map* $j : \pi \longrightarrow \overline{C}$ *is injective and there is an exact sequence of G-modules*

$$0 \longrightarrow \pi \xrightarrow{j} \overline{C} \xrightarrow{d} N \longrightarrow 0$$

*in which* $\bar{C}$ *is the free* G*-module on the elements* $\bar{v}(\rho) = \rho[C, C]$ , $\rho \in R$ *, and* d *is given by* $d(\rho[C, C]) = w(\rho)[N, N]$, $\rho \in R$ .

*Proof.* Since F is free, so also is N , and so the surjection $\partial': C \longrightarrow N$ has a section. □

From now on, we call $\pi$ the *module of identities* for $P = (X; R, w)$ , or the *module of identities among the relations of* P .

## 4. The Associated chain complex

Given a presentation $P = (X; R, w)$ of a group G , there is a standard way of constructing a chain complex of free (right) G-modules

$$C(P): C_2(P) \xrightarrow{d_2} C_1(P) \xrightarrow{d_1} C_0(P)$$

in which

$$C_0(P) = \mathbb{Z}G ,$$

$$C_1(P) = \bigoplus_X \mathbb{Z}G ,$$

$$C_2(P) = \bigoplus_R \mathbb{Z}G$$

with bases (as $\mathbb{Z}G$-modules) respectively 1; $e^1_x$ for $x \in X$ ; and $e^2_\rho$ for $\rho \in R$ . Let F be the free group on X and let the projections $F \longrightarrow G$ , $\mathbb{Z}F \longrightarrow \mathbb{Z}G$ determined by the presentation both be denoted by $\phi$ . Then the boundaries are given by

$$d_1(e^1_x) = 1 - \phi x , \quad x \in X$$

$$d_2(e^2_\rho) = \sum_X e^1_x . \phi(\partial r/\partial x) , \quad \rho \in R$$

where $\partial r/\partial x$ is the element of $\mathbb{Z}F$ known as the *Reidemeister-Fox derivative* of $r = w(\rho)$ . It is computed as follows (see for example [C-F], [Bi]) .

First recall that a *derivation* f from a group $\Gamma$ to a (right) $\Gamma$-module M is a function $f: \Gamma \to M$ satisfying

$$f(uv) = f(u).v + f(v), \quad u, v \in \Gamma \qquad (4.1).$$

This implies that $f(1) = 0$ and that

$$f(u^{-1}) = -f(u).u^{-1}, \quad u \in \Gamma \qquad (4.2).$$

From (4.1) it follows that if $u \in \Gamma$ can be written as $u = y_1 \dots y_n$ then

$$f(u) = f(y_1)\, y_2 \dots y_n + f(y_2)\, y_3 \dots y_n + \dots + f(y_n) \qquad (4.3).$$

It follows from this and (4.2) that if $\Gamma = F$, the free group on $X$, then a derivation $f$ on $F$ may be computed from the values $f(x)$, $x \in X$. It is also not hard to prove the converse, that given the values $f(x)$, $x \in X$, the formulae (4.2) and (4.3) (with $y_i \in X \cup X^{-1}$) determine uniquely a derivation $f$. (The neatest proof relies on the fact that the derivations $F \longrightarrow M$ are bijective with the right inverses of the projection $M \tilde{\times} F \longrightarrow F$ of the semi-direct product of $M$ and $F$, cf. for example [Hi-St], p.196.)

It follows that for any $x$ in $X$ there is a unique derivation $F \longrightarrow \mathbb{Z}F$ whose value on a basis element $x' \in X$ is $\delta_{xx'}$ (the Kronecker delta). This derivation is written $\partial/\partial x$. We can now give the formula for $d_2$ as follows: suppose $w(\rho) = y_1 \dots y_n \in F$ where $y_i \in X \cup X^{-1}$, $i = 1, \dots, n$; then

$$d_2(e^2_\rho) = \sum_{i=1}^{n} \hat{y}_i \cdot \phi(y_{i+1}\, y_{i+2} \dots y_n) \qquad (4.4)$$

where if $y \in X \cup X^{-1}$,

$$\hat{y} = \begin{cases} e^1_x & \text{if } y = x \in X, \\ -e^1_x(\phi x)^{-1} & \text{if } y^{-1} = x \in X. \end{cases}$$

We shall use this formula for $d_2$ in §5.

There is another way of expressing this formula. For any group $\Gamma$ the functor $\mathrm{Der}(\Gamma, -)$ (of derivations from $\Gamma$ to $-$) is represented by the augmentation ideal $I\Gamma$, and any derivation $f\colon \Gamma \longrightarrow M$ is uniquely the composite of a homomorphism $f^*\colon I\Gamma \longrightarrow M$ of $\Gamma$-modules and the derivation $\Gamma \longrightarrow I\Gamma$, $u \longmapsto 1 - u$ (see for example [Hi-St], p.194). If $\Gamma = F$ as above, then $I\Gamma$ is the free $\Gamma$-module on the elements $1 - x$, $x \in X$ (*loc. cit.* p.196), and so one may identify $\bigoplus_X \mathbb{Z}F$ and $IF$ by the rule $e^1_x \longmapsto 1 - x$. So one has an identification

$$\bigoplus_X \mathbb{Z}G \longrightarrow IF \otimes_F \mathbb{Z}G$$

given by $e^1_x \longmapsto (1 - x) \otimes 1$. With this identification $d_2$ may be described simply as

$$d_2(e^2_\rho) = (1 - r) \otimes 1 \in \mathrm{IF} \otimes_F \mathbb{Z}G\,, \quad \text{where} \quad r = w(\rho).$$

This description is often convenient in homological algebra.

PROPOSITION 8. *The module* $\pi$ *of identities for* $P = (X; R, w)$ *is isomorphic to the second homology module* $H_2(C(P))$ *, i.e. to the kernel of* $d_2$ .

*Proof.* In the previous section we have constructed an exact sequence of G-modules

$$0 \longrightarrow \pi \longrightarrow \bar{C} \xrightarrow{d} \bar{N} \longrightarrow 0\,.$$

We shall prove later (Corollary 1 to Proposition 9) that the rule $r \longmapsto d_2(e^2_\rho)$ $(r = w\rho)$ induces an injection $i: \bar{N} \longrightarrow \bigoplus_X \mathbb{Z}G$ ; an algebraic proof of this, using the latter form of $d_2$ , is given for example on p.199 of [Hi-St]. So we have a commutative diagram

$$\begin{array}{ccc} \bar{C} & \xrightarrow{\cong} & C_2(P) \\ {\scriptstyle d}\big\downarrow & & \big\downarrow{\scriptstyle d_2} \\ \bar{N} & \xrightarrow{\;i\;} & C_1(P) \end{array}$$

with $i$ injective, and hence $\pi$ is isomorphic to the kernel of $d_2$ . □

## 5. Relation with 2-dimensional CW-complexes

Let $K$ be a connected CW-complex of dimension 2. Shrinking a tree in $K^1$ to a point does not change the homotopy type of $K$, and so we assume that $K$ has only one vertex, say $a$ . Then the fundamental group $G = \pi_1(K, a)$ has a presentation $(X; R, w)$ such that the elements $x$ of $X$ are bijective with the 1-cells $e^1_x$ of $K$ ; the elements $\rho$ of $R$ are bijective with the 2-cells $e^2_\rho$ of $K$ ; and the relators $r = w(\rho)$ , $\rho \in R$ are determined up to conjugacy by the attaching maps $f_\rho: S^1 \longrightarrow K^1$ of the $e^2_\rho$'s .

Conversely, given a presentation $P = (X; R, w)$ of a group $G$ , one can form a CW-complex $K = K(P)$ with one vertex $a$ ; a 1-cell $e^1_x$ for each element $x$ of $X$ (so that $\pi_1(K^1, a)$ is the free group $F$ on $X$) , and a 2-cell $e^2_\rho$ for each element $\rho$ of $R$ , attached by a representative of the relator $r = w\rho$ in $F$ . The homotopy type of $K(P)$ (and in fact the simple homotopy type [S1, Wr]) is independent of the choice of representative attaching maps for the 2-cells. Note also that for $K(P)$, the attaching maps $f_\rho$ preserve the base point, i.e. $f_\rho(1) = a$. We call $K(P)$ the *geometric realisation* of $P$.

We shall show how to identify the chain complex $C(P)$ of the presentation with the cellular chain complex of the universal cover $\tilde{K}$ of $K = K(P)$. For this, recall that the cells of $\tilde{K}$ have characteristic maps which are precisely the lifts of the characteristic maps of the cells of $K$. This gives a convenient notation for the cells of $\tilde{K}$ as follows.

The set $\tilde{K}^0$ of vertices of $\tilde{K}$ is simply $G = \pi_1(K, a)$. The 1-cells of $\tilde{K}$ are bijective with $X \times G$ and so are written $e^1_{(x,\ g)}$, $(x, g) \in X \times G$, and $e^1_{(x,\ g)}$ joins $(\phi x)g$ to $g$, where $\phi\colon F \longrightarrow G$ is the projection. We write the edge path along $e^1_{(x,\ g)}$ as $(x, g)$ and its inverse as $(x, g)^{-1} = (x^{-1}, (\phi x)\, g)$. The 2-cells of $\tilde{K}$ are bijective with $R \times G$ and so are written as $e^2_{(\rho,\ g)}$, $(\rho, g) \in R \times G$, this cell being attached by a map $f^g_\rho\colon S^1 \longrightarrow \tilde{K}^1$ lifting the attaching map $f_\rho$ of $e^2_\rho$. Suppose the class of $f_\rho$ in $F = \pi_1(K^1, a)$ is $r = y_1 \dots y_n$ where $y_i \in X \cup X^{-1}$, $i = 1, \dots, n$. Then by the uniqueness of path-lifting, the class of $f^g_\rho$ in $\pi_1(\tilde{K}_1, g)$ contains the edge path

$$\left.\begin{array}{l} \qquad\qquad (y_1, g_1)\ (y_2, g_2) \dots (y_n, g_n) \\ \text{where} \\ g_1 = \phi(y_2 \dots y_n)\, g,\ g_2 = \phi(y_3, \dots, y_n)\, g, \dots, g_n = g\ . \end{array}\right\} \qquad (5.1)$$

PROPOSITION 9. *The cellular chain complex* $(C_*(\tilde{K}), \partial)$ *of the universal cover* $\tilde{K}$ *of* $K = K(P)$ *is G-isomorphic to the chain complex* $C(P)$ *associated to the presentation* $P$.

*Proof*. The cellular chain group $C_i(\tilde{K}) = H_i(\tilde{K}^i, \tilde{K}^{i-1})$ is the free abelian group on the i-cells of $\tilde{K}$, and so has a base which can be identified with $R \times G$ if $i = 2$, with $X \times G$ if $i = 1$ and with $G$ if $i = 0$. Since $e^1_{(x,\ g)}$ joins $(\phi x)g$ to $g$ we have

$$\partial_1(x, g) = (1 - \phi x)g\ , \quad (x, g) \in X \times G\ .$$

Suppose $(\rho, g) \in R \times G$ and $w\rho = y_1 \dots y_n$ where $y_i \in X \cup X^{-1}$. Then by the description above of $f^g_\rho$

$$\partial_2(\rho, g) = (y_1, g_1) + (y_2, g_2) + \ldots + (y_n, g_n)$$

where the $g_i$ are given by (5.1). It follows from this and (4.4) that the map $C_*(\tilde{K}) \longrightarrow C(P)$ given on the basis elements by $(\rho, g) \longmapsto e^2_\rho \cdot g$ in dimension 2 ; $(x, g) \longmapsto e^1_x \cdot g$ in dimension 1 , and $g \longmapsto g$ in dimension 0 , is an isomorphism. □

COROLLARY 1. *If* $P = (X; R, w)$ *is a presentation of a group* $G$ *and* $\bar{N}$ *is the relation module of* $P$, *then the rule* $r \longmapsto \Sigma e^1_x\ \phi(\partial r/\partial x)$ *induces an injection* $i: \bar{N} \longrightarrow \oplus_X \mathbb{Z}G$ .

*Proof.* We have identifications

$$\oplus_X \mathbb{Z}G = C_1(\tilde{K}) = H_1(\tilde{K}^1, \tilde{K}^0) .$$

The homology exact sequence of the pair $(\tilde{K}^1, \tilde{K}^0)$ gives an injection $j: H_1(\tilde{K}^1) \longrightarrow H_1(\tilde{K}^1, \tilde{K}^0)$ . The covering projection $p: \tilde{K} \longrightarrow K$ induces an isomorphism $\pi_1(\tilde{K}^1, 1) \longrightarrow N$ which maps the class of the edge path $(y_1, g_1) \ldots (y_n, g_n)$ (as in (5.1)) to $r = y_1 \ldots y_n$, $y_i \in X \cup X^{-1}$ . The Hurewicz map $\pi_1(\tilde{K}^1, 1) \longrightarrow H_1(\tilde{K}^1)$ thus induces an isomorphism $\bar{N} \longrightarrow H_1(\tilde{K}^1)$ and the composite of $j$ with this isomorphism is the map $i$ . □

Corollary 1 may also be proved using the methods of §3.1 of [G1] . Given a group $G$ , and short exact sequence $1 \longrightarrow N \longrightarrow F \longrightarrow G \longrightarrow 1$ with $F$ free, Gruenberg constructs a free G-resolution of $\mathbb{Z}$

$$\ldots \to \mathsf{N}^2/\mathsf{N}^3 \to \mathsf{FN}/\mathsf{FN}^2 \to \mathsf{N}/\mathsf{N}^2 \to \mathsf{F}/\mathsf{FN} \to \mathbb{Z}G \to \mathbb{Z} \to 0$$

in which $\mathsf{F}$ is the augmentation ideal $IF$ of $F$ and $\mathsf{N}$ is the kernel of the induced map $\mathbb{Z}F \longrightarrow \mathbb{Z}G$ . If $F$ is free on $X$ , and $N$ is the free group on $V$ , then $\mathsf{N}/\mathsf{N}^2$ and $\mathsf{F}/\mathsf{FN}$ are free G-modules on the cosets of the elements $1 - v$ , $v \in V$ and $1 - x$ , $x \in X$ , respectively. Thus $\mathsf{F}/\mathsf{FN}$ is isomorphic to our $C_1(P)$ but in general $\mathsf{N}/\mathsf{N}^2$ is not isomorphic to $C_2(P)$ . However the map $N \longrightarrow \mathsf{N} \longrightarrow \mathsf{N}/\mathsf{FN}$ (which sends $n \in N$ to the coset of $1 - n$) induces an isomorphism of abelian groups $\bar{N} \longrightarrow \mathsf{N}/\mathsf{FN}$ , and the map $\mathsf{N}/\mathsf{FN} \longrightarrow \mathsf{F}/\mathsf{FN}$ is an injection.

COROLLARY 2. *Let* $P = (X; R, w)$ *be a presentation of a group* $G$ , *and let* $K = K(P)$ *be its geometric realisation. Then the module* $\pi$ *of identities for* $P$ *is naturally isomorphic to the second homology group* $H_2(\tilde{K})$ *of the universal cover* $\tilde{K}$ *of* $K$ , *and hence also to* $\pi_2(K)$ , *the second homotopy group of* $K$ .

*Proof.* The first assertion is immediate from Propositions 7 and 8, while the second follows from the Hurewicz theorem, since $\pi_2(\tilde{K}) \cong \pi_2(K)$ . □

The above description of the module of identities as an absolute homotopy group can be extended to a description of the free crossed F-module C of the presentation as a relative homotopy group. The history of this description is as follows.

In his 1941 paper [Wh1] , Whitehead attempted an algebraic description of the second homotopy group $\pi_2(K)$ of a space $K = L \cup \{e^2_\rho\}_{\rho \in R}$ obtained by attaching 2-cells $e^2_\rho$ to a path-connected space $L$ . He reformulated these results in [Wh2] as a precise algebraic description of the group $\pi_2(K, L)$ and also noted that if $L$ is a 1-dimensional complex, then his description of $\pi_2(K)$ returned to previous results of Reidemeister [Re1] (see Corollary 2 above).

A fundamental observation in [Wh2] is that if $(Z, Y)$ is any based pair of spaces, then the second relative homotopy group $\pi_2(Z, Y)$ has an action of $\pi_1(Y)$ so that with the boundary map $\partial: \pi_2(Z, Y) \longrightarrow \pi_1(Y)$ , the rules (CM1), (CM2) hold. (For proofs of these rules, see for example [Hi] p.39 or [W] .) This led Whitehead to the definition of crossed module [Wh3] .

For the particular pair $(K, L)$ , where $K = L \cup \{e^2_\rho\}_{\rho \in R}$ as above, we can obtain elements $a_\rho \in \pi_2(K, L)$ , given the characteristic maps $h_\rho: (E^2, S^1) \longrightarrow (K, L)$ of the 2-cells $e^2_\rho$ together with a choice of paths in $L$ , one for each $\rho$ , joining $h_\rho(1)$ to the base point of $L$ . We can now state a theorem from [Wh3] .

THEOREM 10. *The crossed* $\pi_1 L$*-module* $\pi_2(K, L)$ , *where* $K = L \cup \{e^2_\rho\}_{\rho \in R}$ , *is free on the elements* $a_\rho$ , $\rho \in R$ .

COROLLARY. *If* $K = K(P)$ *is the geometric realisation of a presentation* $P$ , *then the free crossed* F*-module* $(C, \partial)$ *of* $P$ *is isomorphic, given the identification* $F = \pi_1 K^1$ , *to the crossed* $\pi_1 K^1$*-module* $(\pi_2(K^2, K^1), \partial)$; *in particular, the module* $\pi$ *of identities of* $P$ *is isomorphic to* $\pi_2(K)$ . □

Whitehead's proof of Theorem 10 uses methods of transversality and knot theory - an exposition of this proof is given in [Br1]. The theorem is also a special case of the generalised Seifert-van Kampen theorem of [B-H2], sketched in [Br2] in this volume. In the case $L$ is a 1-dimensional CW-complex, a short proof was given in [C1] as an application of the relative Hurewicz theorem, and this method has been extended to the general case in [R]. For completeness, we give another proof here of the special case, without using the Hurewicz theorem.

*Proof of Theorem* 10 *for the case* $L = K^1$. Clearly we may assume $K$ is of the form $K(P)$ for a presentation $P = (X; R, w)$ of a group $G$. Let $(C, \partial)$ be the free crossed $F$-module on $w: R \to F$, as in §3. Then there is a unique homomorphism $\phi: C \longrightarrow \pi_2(K^2, K^1)$ of crossed $F$-modules such that $\phi\rho = a_\rho$, $\rho \in R$. So we obtain a commutative diagram

$$\begin{array}{ccccccccc}
0 & \longrightarrow & \pi & \xrightarrow{\ i\ } & C & \longrightarrow & N & \longrightarrow & 1 \\
 & & \downarrow{\scriptstyle \phi'} & & \downarrow{\scriptstyle \phi} & & \| & & \\
0 & \longrightarrow & \pi_2 K & \xrightarrow[j]{} & \pi_2(K^2, K^1) & \longrightarrow & N & \longrightarrow & 1 .
\end{array}$$

That $\phi$ is surjective is fairly easily proved by a general position argument which will be given in §10 below (cf. also [W] and [Br1]). The more difficult part is to prove $\phi$ injective.

We have isomorphisms given previously

$$\overline{C} \cong \bigoplus_R \mathbb{Z}G \cong H_2(\tilde{K}, \tilde{K}^1) .$$

Thus the Hurewicz map $\pi_2(K, K^1) \longrightarrow H_2(\tilde{K}, \tilde{K}^1)$ determines a map $\psi: \pi_2(K, K^1) \longrightarrow \overline{C}$ such that $\psi(a_\rho) = e^2_\rho$, $\rho \in R$. Hence, the abelianised maps $\overline{\psi}$, $\overline{\phi}$ satisfy $\overline{\psi}\overline{\phi} = 1$. Let $q: C \longrightarrow \overline{C}$, $q': \pi_2(K^2, K^1) \longrightarrow \overline{\pi}_2(K^2, K^1)$ be the abelianising maps. Then

$$\overline{\psi}\, q'\, \phi i = \overline{\psi}\, \overline{\phi}\, q\, i = q\, i$$

which is injective by Proposition 4. Hence $\phi i$ is injective and so $\phi'$ is injective. By the 5-lemma applied to the above diagram, $\phi$ is injective. □

REMARKS 1. These results do give precise information on $\pi_2(K)$, where $K$ is a 2-complex, or equivalently, on the module of identities for a presentation $(X; R, w)$; however they are not so easy to interpret in practice. Quite a lot of information is known on relation modules, particularly for abelian groups [G2, S-D, We]. See [D4, Hu2, G-R] for some results on $\pi_2(K)$.

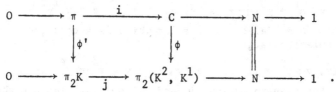

2. Dyer-Vasquez [D-V] have a different method of constructing a complex $K(P)$ of a one-relator presentation $P = (X; r)$ of a group $G$. If $r$ is not a proper power, they proceed as above. However if $r = z^q$, where $q > 1$ is maximal, then they attach to $K^1$ not a 2-cell but an Eilenberg-MacLane space $K(Z_q, 1) = S^1 \cup_q e^2 \cup e^3 \cup \ldots$ by means of a map $S^1 \longrightarrow K^1$ representing $z$. This yields an Eilenberg-MacLane space $K(G, 1)$.

3. If $P$, $P'$ are two presentations of a group $G$, then (see for example [C-F], [J]) $P$ can be transformed to $P'$ by a sequence of Tietze transformations, which are

I (and I') Add (delete) a generator and relation which expresses that generator as a word in the other generators, e.g.

$$(x, y; x^2 = y^3) \longmapsto (x, y, z; x^2 = y^3, z = x^{-1}y^2x)$$

II (and II') Add (delete) a relation which is a consequence of the other relations e.g.

$$(x, y; x^2 = y^3) \longmapsto (x, y; x^2 = y^3, x^4 = y^6).$$

Instead of the transformations II and II' one can also use the transformations (cf. [D4, Me1, S1, Wa]):

II a Replace a relator $r$ by $rw^{-1}sw$ or $rw^{-1}s^{-1}w$ where $s$ is another relator and $w$ is an arbitrary word in the generators.

II b (and II b') Add (delete) the relation "1 = 1" (the corresponding relator is the identity).

A transformation II a is a product of a transformation II and a transformation II'. A transformation II (II') can easily be written as a product of transformations II a and II b(II b').

Sieradski in [S1] calls presentations equivalent under the use of operations I, I' and II a *combinatorially equivalent*. (Actually, he used a different, but equivalent, set of operations.) There are a number of problems in this area. The following is taken from Problem 5.1 of [K].

Let $K(P)$, $K(P')$ be the geometric realisations of two finite presentations $P$, $P'$ of a group $G$. Assume $P$, $P'$ have the same *deficiency* (= number of generators - number of relators). Consider the assertions:

A) $K(P) \simeq K(P')$ (homotopy equivalence)

B) $K(P) \curlywedge K(P')$ (simple homotopy equivalence)

C) $K(P) \underset{3}{\curlywedge} K(P')$ (simple homotopy equivalence by moves of dimension $\leq 3$)

D) $P$ is combinatorially equivalent to $P'$ .

Then $D \Leftrightarrow C \Rightarrow B \Rightarrow A$ [Wr] (see also [Me 1, 2]) . It is not known what other relations hold in general. For more discussion of this and other problems on 2-complexes, see also [Wa2] .

The main results of [S1] and [Me1] give presentations of finitely generated abelian groups which are not combinatorially equivalent. The simplest example is the two presentations $(x, y; x^5, y^5, [x, y])$ and $(x, y; x^5, y^5, [x^2, y])$ of $\mathbb{Z}_5 \times \mathbb{Z}_5$. The proof involved considering the map $C(P) \longrightarrow C(P')$ of chain complexes induced by a combinatorial equivalence from $P$ to $P'$ . [Me1] also considers coarser equivalences (allowing also permutations of the generators).

The presentation $P = (x, y; x^2y^{-3}, 1)$ of the trefoil group $G$ has its module $\pi$ of identities isomorphic to $\mathbb{Z}G$ . In [D4], Dunwoody constructs for $G$ another presentation $P'$ , with two generators and two relators, for which the module $\pi'$ of identities is not free. Since $\pi$ and $\pi'$ are not isomorphic, the spaces $K(P)$, $K(P')$ are not of the same homotopy type. However, he also proves that $\pi \oplus \mathbb{Z}G \cong \pi' \oplus \mathbb{Z}G$ , and that $K(P) \vee S^2$ and $K(P') \vee S^2$ are of the same homotopy type. Thus $\pi'$ is projective and stably free. Other examples of non-free projective modules over $\mathbb{Z}G$ where $G$ is torsion free are given in [Be-Du] (for $G$ the trefoil group made matabelian), but it is not known if these are isomorphic to the second homotopy module of a 2-complex. Also by the Corollary on p.139 of [Wald] , $\tilde{K}_0(\mathbb{Z}G) = 0$ for $G$ in a large class which includes (by Theorem 17.5 *op. cit.*) all poly-$\mathbb{Z}$-groups, all torsion free one-relator groups, and fundamental groups of compact, orientable 3-manifolds which are sufficiently large (this includes for example the trefoil group). Hence projective modules over such groups are stably free. Many examples of non-free projective modules over the rational group ring of a torsion free group are known [Le] , but the existence of these does not imply such examples exist over the integral group ring. For further discussion of related problems, see [Ba] .

## 6. Peiffer transformations

As before, let $P = (X; R, w)$ be a presentation of a group $G$ . Let $F$ be the free group on $X$ . The aim of this section is to give a more combinatorial description of the free crossed $F$-module $C$ on $w : R \longrightarrow F$ . This description, which is essentially due to Peiffer [Pe], will be useful later.

Recall that we considered in §1 the free group $H$ on the set

$Y = R \times F$ , with elements of $Y$ written $a = \rho^u$, $\rho \in R$, $u \in F$ . The combinatorial description of $C$ uses operations on words rather than on elements of $H$ ; a word in the elements of $Y$ is written as an n-tuple

$$\underset{\sim}{y} = (a_1, \ldots, a_n), \text{ where } a_i = (\rho_i^{u_i})^{\varepsilon_i}, \varepsilon_i = \pm 1, \rho_i \in R, u_i \in F,$$

for some $n \geq 0$ . We shall refer to such a sequence as a Y-*sequence*, and shall write $\theta\underset{\sim}{y}$ for the product $(\theta a_1) \ldots (\theta a_n)$ in $F$ , where, as in §1, $\theta((\rho^u)^\varepsilon) = u^{-1}(w\rho)^\varepsilon u$ . If $\theta y = 1$ , we call $\underset{\sim}{y}$ an *identity* Y-*sequence* for $P$ . For example,

$$\underset{\sim}{y} = (\rho^u, \rho^{-u}, \sigma^{-v}, \sigma^v, \tau^w, \tau^{-w})$$

where $\sigma, \rho, \tau \in R$ and $u, v, w \in F$ , is an identity Y-sequence, and in this case the corresponding element of $H$ is $1$ .

In §2 we formed the Peiffer group $P$ of the pre-crossed F-module $(H, \theta)$ as the subgroup of $H$ generated by the Peiffer elements $b^{-1}a^{-1}ba^{\theta b}$ for all $a , b \in H$ . This means that if we work mod $P$ in $H$ we have the "crossed commutation" rules

$$ab \equiv ba^{\theta b}, \quad ab \equiv b^{\theta a^{-1}} a \bmod P$$

for all $a, b \in H$ . Further, the Corollary to Proposition 3 shows that these rules for all $a , b$ of $H$ are a consequence of the rules simply for all elements $a , b$ of $Y$ . Such rules, together with the rule $a a^{-1} = 1$ , can be modelled on words in $Y$ , by certain operations which we now explain.

Peiffer operations on Y-sequences:

(i) An *elementary Peiffer exchange* replaces an adjacent pair $(a, b)$ in a Y-sequence by either $(b, a^{\theta b})$ or $(b^{\theta a^{-1}}, a)$ . A *Peiffer exchange* is a sequence of elementary Peiffer exchanges; we often abbreviate "Peiffer exchange" to "exchange".

(ii) A *Peiffer deletion* deletes an adjacent pair $(a, a^{-1})$ in a Y-sequence. A *Peiffer collapse* is a sequence of exchanges and Peiffer deletions, in some order.

(iii) A *Peiffer insertion* is the inverse of a Peiffer deletion, and a *Peiffer expansion* is the inverse of a Peiffer collapse.

(iv) A *Peiffer equivalence* is a sequence of Peiffer collapses and Peiffer expansions, in some order.

REMARK. Operations of this kind are considered in [Pe] . A number of authors have used some coarser operations which we shall discuss later and call simply *collapses*, *expansions* and *equivalences*.

Given a Y-sequence $\underset{\sim}{y}$ , we obtain an element $\psi y$ of the free group H on Y by forming the product in H of the components of $\underset{\sim}{y}$ . By the construction of free groups, $\psi y = \psi z$ if and only if $\underset{\sim}{z}$ can be obtained from $\underset{\sim}{y}$ by a sequence of Peiffer deletions and Peiffer insertions, in some order. The definitions of the Peiffer group P and the free crossed F-module $C = H/P$ give immediately:

PROPOSITION 11. *Two* Y-*sequences have the same image in* $C = H/P$ *if and only if they are Peiffer equivalent.*

For later use, we also give a simple but useful observation on exchange operations.

PROPOSITION 12. *If a* Y-*sequence* $\underset{\sim}{z}$ *is obtained from a* Y-*sequence* $\underset{\sim}{y} = (a_1, \ldots, a_m)$ *by Peiffer exchanges, then each component of* $\underset{\sim}{z}$ *is of the form*

$$a_i^{v_i} \quad \textit{for} \quad v_i \in gp\{\theta a_1, \ldots, \theta a_n\} ;$$

*in particular,* $v_i$ *belongs to* N *, the normal closure of the relators.*

The proof is clear from the definition of Peiffer exchange. Note that each $a_i$ is of the form $u_i^{-1} r_i^{\varepsilon_i} u_i$ , $r_i \in R$ , $u_i \in F$ , and so the subgroup of F generated by the $\theta a_1, \ldots, \theta a_n$ is a subgroup of N .

The Peiffer equivalences turn out to be particularly relevant for a class of presentations called 'aspherical' (§7). The groups of such presentations are torsion-free. A wider class of groups and presentations can be discussed using a larger class of operations than the Peiffer equivalences - for example, in this way one studies the 'combinatorially aspherical' presentations (§8); these determine groups among which are one-relator groups, most Fuchsian groups, and many others. The definition of this wider class of operations is as follows.

Operations on Y-sequences:

(i) *Exchanges* will be the Peiffer exchanges as above.

(ii) A *deletion* is a deletion of an adjacent pair (a, b) in a Y-sequence in case $(\theta a)(\theta b) = 1$ in F . A *collapse* is a sequence of exchanges and deletions in some order.

(iii) An *insertion* is the inverse of a deletion and an *expansion* is the inverse of a collapse. (But note that to insert (a, b) in a Y-sequence we must have not only $(\theta a)(\theta b) = 1$ but also $a , b \in Y \cup Y^{-1}$ , so that we still have a Y-sequence.)

(iv) An *equivalence* of Y-sequences is a sequence of collapses and expansions, in some order.

Clearly Peiffer equivalence implies equivalence; it is useful to know when the converse holds.

We say that the presentation $P = (X; R, w)$ is *redundant* if (i) there is a $\tau$ in $R$ such that $w\tau = 1$, or (ii) there are $\rho$, $\sigma$ in $R$ such that $\rho \neq \sigma$ but $w\rho$ is conjugate to $w\sigma$ or to $w\sigma^{-1}$. (If $P$ is not redundant, it is *irredundant*.) If $P$ is redundant, we can find $a$, $b$ in $Y \cup Y^{-1}$ such that $(\theta a)(\theta b) = 1$ but $b \neq a^{-1}$; so in this case, an insertion or deletion for a Y-sequence need not be a Peiffer insertion or Peiffer deletion.

We say the presentation $P$ is *primary* if for all $\rho \in R$, $w\rho$ is not a proper power. If this does not hold, then, by Example 2 of §1, we can again find an insertion (or deletion) which is not a Peiffer insertion (or Peiffer deletion).

PROPOSITION 13. *Let* $P = (X; R, w)$ *be a presentation which is irredundant and primary. Then any deletion (insertion) has the same effect as a suitable Peiffer deletion (Peiffer insertion), combined with a sequence of elementary exchanges.*

*Proof.* Suppose given $a = (\rho^u)^\varepsilon$, $b = (\sigma^v)^\eta$, elements of $Y \cup Y^{-1}$, such that $(\theta a)(\theta b) = 1$. Let $r = w\rho$, $s = w\sigma$. With this notation, we have the following lemma.

LEMMA. *If* $P$ *is irredundant, then* $\rho = \sigma$, $r = s$, $\varepsilon + \eta = 0$ *and for some* $m \in \mathbb{Z}$, $uv^{-1} = z^m$, *where* $z$ *is the root of* $r$.

The proof of the lemma is easy. We are given $r^\varepsilon = uv^{-1}s^{-\eta}vu^{-1}$. By irredundancy, $\rho = \sigma$, $r = s$ and hence (since $r \neq 1$), $\varepsilon + \eta = 0$. So $uv^{-1}$ centralises $r$, which implies the lemma.

Since also $P$ is primary, we have further that $z = r$. So

$$a = (\rho^u)^\varepsilon, \quad b = (\rho^v)^{-\varepsilon} \quad \text{with} \quad \varepsilon = \pm 1, \quad uv^{-1} = r^m .$$

We now do an elementary exchange of $(a, b)$ to $(a_1, b_1)$ say, where $a_1 = (\rho^{u_1})^{-\varepsilon}$, $b_1 = (\rho^{v_1})^\varepsilon$. If $|m + \varepsilon| < |m|$, we use here the elementary exchange $(a, b) \sim (b^{\theta a^{-1}}, a)$ and obtain easily that $u_1v_1^{-1} = r^{-(m+\varepsilon)}$. If $|m - \varepsilon| < |m|$, we use the elementary exchange $(a, b) \sim (b, a^{\theta b})$ and obtain that

$u_1v_1^{-1} = r^{-(m-\varepsilon)}$ . Hence a sequence of $|m|$ elementary exchanges carries $(a, b)$ to an identity sequence $(a_m, b_m)$ with $a_m = b_m^{-1}$ (as elements of $Y \cup Y^{-1}$) . This clearly implies the assertion. □

COROLLARY. *Let* $P = (X; R, w)$ *be a presentation which is irredundant and primary. Then two* Y*-sequences determine the same element of the free crossed* F*-module* $C = H/P$ *if and only if they are equivalent.* □

REMARK. Let $P = (X; R, w)$ be a presentation and let $R = w(R)$ , $F = F(X)$ as usual. It is common in the literature to consider not the Y-sequences in the above (where $Y = R \times F$ ) but what we could call the $R^F$-sequences $\underset{\sim}{p} = (p_1, \ldots, p_n)$ where each $p_i$ is a conjugate of a relator or its inverse, so that each $p_i$ is an element of $N = N(R)$ . If $\underset{\sim}{y} = (a_1, \ldots, a_n)$ is a Y-sequence, then $\theta'\underset{\sim}{y} = (\theta a_1, \ldots, \theta a_n)$ is an $R^F$-sequence. We say $\underset{\sim}{p}$ is an *identity* $R^F$-sequence if $p_1 \ldots p_n = 1$ in $F$ . Clearly $\underset{\sim}{y}$ is an identity Y-sequence if and only if $\theta'\underset{\sim}{y}$ is an identity $R^F$-sequence. It is these identity $R^F$-sequences which are considered in [L-S] and [C-C-H] . The operations on Y-sequences can be modelled in $R^F$-sequences. The *elementary exchanges* replace an adjacent $(p, q)$ in $\underset{\sim}{p}$ by $(q, p^q)$ or $(q^{p^{-1}}, p)$ ; these are called *exchanges* in [C-C-H] and *Peiffer transformations of the first kind* in [L-S] . The *deletions* or *insertions* delete or insert an adjacent $(p, p^{-1})$ . (The deletions are called *Peiffer transformations of the second kind* in [L-S], and insertions are not considered. Both operations are considered in [C-C-H].) *Equivalences* of $R^F$-sequences are composites of exchanges, deletions and insertions, in some order. Clearly the map $\theta'$ from Y-sequences to $R^F$-sequences induces a bijection of equivalence classes. However, for $R^F$-sequences there is no notion corresponding to our Peiffer equivalence, and so in general we do not recover the free crossed F-module $(C, \partial)$ of $P$ from the $R^F$-sequences. Nonetheless by the last Corollary, we may recover $C$ , and hence the module $\pi$ of identities, from the $R^F$-sequences if $P$ is irredundant and primary.

## 7. Aspherical 2-complexes and aspherical presentations

A topological space $X$ is *aspherical* if it is connected and $\pi_i X = 0$ for $i > 1$ . Thus for such $X$ the significant homotopy invariant is the fundamental group $\pi_1 X$ , and for aspherical CW-complexes $X$ the fundamental group determines the homotopy type of $X$ . (See for example [W].) If $K$ is a connected 2-dimensional CW-complex, then $K$ is aspherical if and only if $\pi_2 K = 0$ .

PROPOSITION 14. *Let* $K = K(P)$ *be the geometric realisation of a presentation* $P = (X; R, w)$ . *Then the following are equivalent.*

(i) *The 2-complex* $K$ *is aspherical, i.e.* $\pi_2 K = 0$ .

(ii) *The module* $\pi$ *of identities for* $P$ *is zero.*

(iii) *The relation module* $\bar{N}_-$ *of* $P$ *is the free module on the induced map* $\bar{w}: R \longrightarrow \bar{N}$ .

(iv) *Any identity* Y-*sequence for* $P$ *is Peiffer equivalent to the empty sequence.*

*Proof.* That (i) and (ii) are equivalent is immediate from Corollary 2 of Proposition 9. That (ii) and (iii) are equivalent follows from the Corollary to Proposition 7. Finally, the equivalence of (ii) and (iv) follows from Proposition 11. □

We now follow [T, S1, C-C-H] in calling a presentation $P$ *aspherical* if $\pi_2 K(P) = 0$ , i.e. if any of the equivalent properties of Proposition 14 hold. There is another useful condition for $P$ to be aspherical.

PROPOSITION 15. *A presentation* $P = (X; R, w)$ *is aspherical if and only if* $P$ *is irredundant and primary, and any identity* Y-*sequence for* $P$ *is equivalent to the empty sequence* $\emptyset$ .

*Proof.* Suppose first that $P$ is aspherical. We prove that $P$ is irredundant and primary.

Let $\rho \in R$ . Since $\pi = 0$ , $d_2 e^2_\rho \neq 0$ and so $w\rho \neq 1$ . Let $\sigma \in R$ and suppose $r = w\rho$ is conjugate to $s = w\sigma$ , i.e. $r = u^{-1}su$ for some $u \in F$ . Then the elements $\rho$ , $\sigma^u$ of the free crossed F-module $C$ of $P$ satisfy $\partial\rho = \partial\sigma^u$ . Since $\operatorname{Ker}\partial = \pi = 0$ , we have $\rho = \sigma^u$ in $C$ and so $e^2_\rho = e^2_\sigma \cdot \phi u$ in $C_2(P)$ . By freeness of $C_2(P)$ , $\rho = \sigma$ . A similar proof, with $e^2_\sigma$ replaced by $-e^2_\sigma$ , shows that $r$ cannot equal $u^{-1}s^{-1}u$ .

Suppose now $r = z^q$ where $q \geq 1$ . Then $r = z^{-1}rz$ . The above proof shows that $e^2_\rho = e^2_\rho \cdot \phi z$ and so $\phi z = 1$ . Hence

$z = \partial a$ for some $a \in C$. Then $\partial\rho = \partial a^q$. Since $\mathrm{Ker}\,\partial = 0$, $\rho = a^q$ and therefore in $C_2(P)$, $e^2_\rho$ is divisible by $q$. Since $e^2_\rho$ is a basis element for $C_2(P)$, we have $q = 1$.

This completes the proof that if $P$ is aspherical then it is irredundant and primary. The remaining assertions of the Proposition follow from the Corollary to Proposition 13, and Proposition 14. □

## 8. The identity property

In this section, we present a property of a presentation first described by Lyndon in [L1] and later called the *identity property* by Papakyriakopoulos in [P2]. This property provides a useful characterisation of those presentations which are irredundant and for which any identity sequence is equivalent to the empty sequence. In this section, we abbreviate "Y-sequence" to "sequence".

DEFINITION. Let $P = (X; R, w)$ be a presentation, and let $\underset{\sim}{y} = (a_1, \ldots, a_n)$, where each $a_i = (\rho_i^{u_i})^{\varepsilon_i}$, $\rho_i \in R$, $u_i \in F$, $\varepsilon_i = \pm 1$, be an identity sequence for $P$. We say $\underset{\sim}{y}$ has the *identity property* if the indices $1, \ldots, n$ can be grouped into pairs $(i, j)$ such that $\rho_i = \rho_j$, $\varepsilon_i + \varepsilon_j = 0$ and, if $z_i$ is the root of $r_i = w\rho_i$, then for some $m_i \in \mathbb{Z}$

$$u_i \equiv z_i^{m_i} u_j \bmod N . \qquad (8.1)$$

We say $\underset{\sim}{y}$ has the *primary identity property* if it has the identity property but with (8.1) replaced by

$$u_i \equiv u_j \bmod N . \qquad (8.2)$$

We say $P$ has the *(primary) identity property* if every identity sequence for $P$ has this property.

PROPOSITION 16. *Let* $P$ *be a presentation and let* $\underset{\sim}{y}$ *be an identity sequence for* $P$.

(i) *If* $\underset{\sim}{y}$ *has the identity property then* $\underset{\sim}{y}$ *is equivalent to the empty sequence.*

(ii) *If* $\underset{\sim}{y}$ *has the primary identity property, then* $\underset{\sim}{y}$ *is Peiffer equivalent to the empty sequence.*

*Also, the converse to* (ii) *holds, and the converse to* (i) *holds if* $P$ *is irredundant.*

*Proof.* (i) By exchanges we can transform $\underset{\sim}{y}$ to $\underset{\sim}{z}$ which again has the identity property but with adjacent indices paired. Thus we can write $\underset{\sim}{z} = (b'_1, b_1, b'_2, b_2, \ldots)$ where

$$b'_i = (\rho_i^{u'_i})^{\varepsilon_i}, \quad b_i = (\rho_i^{u_i})^{-\varepsilon_i} \text{ and } u'_i = z_i^{m_i} u_i v_i$$

where $z_i$ centralises $r_i = w\rho_i$, and $v_i$ belongs to $N$. Let $c_i = (\rho_i^{u_i v_i})^{\varepsilon_i}$. Then $\theta c_i = \theta b'_i$, and so by deletions and insertions we can transform $\underset{\sim}{z}$ to $\underset{\sim}{w} = (c_1, b_1, c_2, b_2, \ldots)$. Now $v_i = \partial h_i$ for some $h_i$ in the free crossed module $C$ of $P$. Hence the product

$$w = c_1 b_1 c_2 b_2 \ldots = h_1^{-1} b_1^{-1} h_1 b_1 h_2^{-1} b_2^{-1} h_2 b_2 \ldots \in [C, C] .$$

But $\underset{\sim}{w}$ is an identity sequence, so $\partial w = 1$. By Proposition 4, $w = 1$ in $C$. By Proposition 12, $\underset{\sim}{w}$ is Peiffer equivalent to $\emptyset$. Hence $\underset{\sim}{y}$ is equivalent to $\emptyset$.

(ii) This is proved as for (i), but with $c_i = b'_i$, so that $\underset{\sim}{y}$ is Peiffer equivalent to $\underset{\sim}{w}$ and hence to $\emptyset$.

The converse to (ii) holds, since the empty sequence has the primary identity property, and this property is preserved under Peiffer equivalence. A similar reasoning gives the converse to (i), if $P$ is irredundant. □

We now give another characterisation of the identity property for an identity sequence. This will lead to a characterisation of the identity property for a presentation in terms of the structure of the module of identities or, equivalently, the structure of the relation module.

Let $P = (X; R, w)$ be a presentation of a group $G$. Recall from §§1, 3 and 5 that we have exact sequences

$$1 \longrightarrow E \longrightarrow H \xrightarrow{\theta} F \xrightarrow{\phi} G \longrightarrow 1 ,$$

$$1 \longrightarrow P \longrightarrow H \longrightarrow C \longrightarrow 1 ,$$

$$1 \longrightarrow P \longrightarrow E \longrightarrow \pi \longrightarrow 1 ,$$

$$1 \longrightarrow \pi \longrightarrow C \longrightarrow N \longrightarrow 1 ,$$

$$1 \longrightarrow \pi \longrightarrow \bar{C} \longrightarrow \bar{N} \longrightarrow 1 ,$$

the last of which identifies the module $\pi = E/P$, of identities for $P$, with a submodule of $\bar{C} = C_2(P)$, the free $\mathbb{Z}G$-module with basis $\{e_\rho^2 \quad \rho \in R\}$.

We now construct a module associated with the roots of the

relators of the presentation $P$. Let $\tilde{P}$ be the normal closure in $H$ of the Peiffer group $P$ together with the elements

$$(\rho(\rho^z)^{-1})^u \quad u \in F \ , \ \rho \in R$$

where $z$ is the root of $w\rho$. Notice that $\tilde{P}$ is an F-subgroup of $H$ and lies in $E$. Hence $\tilde{P}/P$ is a G-submodule of $\pi$ and the injection $\pi \to \bar{C}$ identifies $\tilde{P}/P$ with the submodule $M$ of $\bar{C}$ generated by the elements

$$e_\rho^2 \cdot (1 - \phi z) \ , \ \rho \in R \ .$$

We call this submodule the *root module* of $P$.

PROPOSITION 17. *Let* $\underset{\sim}{y} = (a_1, \ldots, a_n)$ *be an identity sequence for the presentation* $P$. *Then the following conditions are equivalent.*

*(i)* $\underset{\sim}{y}$ *has the identity property*

*(ii) The element* $y = a_1 \ldots a_n$ *belongs to* $\tilde{P}$.

*(iii) The image* $\bar{y}$ *of* $y$ *in* $\bar{C}$ *is an element of the root module of* $P$.

*Proof.* That (ii) $\Longleftrightarrow$ (iii) follows from the identification of $\tilde{P}/P$ with $M$.

(i) $\Longrightarrow$ (iii) The pairings given by the identity property imply that $\bar{y}$ is a sum of elements of the form

$$\pm e_\rho^2 \cdot (1 - \phi z^{\pm m}) \ \phi u$$

where $u \in F$, $z$ is the root of $w\rho$ and $m > 0$. The rules

$$(1 - \phi z^m) = (1 - \phi z)(1 + \phi z + \ldots + \phi z^{m-1})$$

$$(1 - \phi z^{-m}) = -(1 - \phi z^m)(\phi z^{-m})$$

now imply that $\bar{y}$ belongs to the root module of $P$.

(ii) $\Longrightarrow$ (i) We are given $y \in \tilde{P}$. Then the image $\bar{y}$ of $y$ in $\pi$ is also the image of an element $p_1 q_1 \ldots p_\ell q_\ell$ of $\tilde{P}$ where the $p_i, q_i$ are respectively of the form $(\rho^u)^\varepsilon$, $(\rho^{zu})^{-\varepsilon}$ where $u \in F$, $\varepsilon = \pm 1$ and $z$ is the root of $w\rho$. So $\underset{\sim}{y}$ is Peiffer equivalent to the sequence $(p_1, q_1, \ldots, p_\ell, q_\ell)$. This sequence has the identity property, and this property is preserved under Peiffer equivalence. Hence $\underset{\sim}{y}$ has the identity property. □

COROLLARY. *Let* $P = (X; R, w)$ *be a presentation of a group* $G$. *Then the following are equivalent:*

*(i)* $P$ *has the identity property.*

*(ii)* $P$ *is irredundant, and each identity sequence for* $P$ *is equivalent to the empty sequence.*

(iii) *P is irredundant, and the root module of P coincides with the module of identities for P .*

(iv) *The relation module $\bar{N}$ of P decomposes, as a $\mathbb{Z}G$-module, into a direct sum of cyclic submodules $\bar{N}_\rho$, $\rho \in R$ , where each $\bar{N}_\rho$ is generated by the image $\bar{r}$ in $\bar{N}$ of the relator $r = w\rho$ , subject to the single relation*

$$\bar{r} \cdot (1 - \phi(z)) = 0 ,$$

*z being the root of r .* □

Since (i) and (iv) clearly imply P irredundant, this is immediate from previous results. Notice that condition (iv) says simply that the map $d_2\colon C_2(P) \longrightarrow C_1(P)$ determines its image $\bar{N}$ as the quotient of $C_2(P)$ by the root module M of P , whence it is clear that (iii) and (iv) are equivalent. Furthermore, it is straightforward to check directly that the identity property implies (iv) .

REMARK 1. Proposition 17 seems to be new. The fact that the identity property for P implies condition (iv) was indicated in [L1] ; that (iv) implies (ii) is due to Huebschmann [Hu3] . In [Hu2] the determination of the module of identities as what we have called the root module is given. A proof that (iv) implies the identity property does not seem to have been given in the literature.

REMARK 2. Various other notions of asphericity for a presentation P are considered in [C-C-H] . These are as follows:

(DA) P is *diagrammatically aspherical* if every identity $R^F$-sequence over P can be transformed to the empty (identity) sequence by collapses.

(SA) P is *singularly aspherical* if it is diagrammatically aspherical, irredundant and primary.

We note in passing that, in view of Propositions 13, 14 and 15, P is singularly aspherical if and only if every identity Y-sequence over P can be transformed to the empty sequence by Peiffer collapses.

For the next two definitions, note that for any presentation $P = (X; R, w)$ , we can find a subpresentation $\hat{P} = (X; \hat{R}, \hat{w})$ of the same group, with $\hat{R}$ contained in R , $\hat{w}$ equal to the restriction of w , and such that $\hat{P}$ is irredundant. We call $\hat{P}$ an *irredundant part* of P .

(CA) P is *combinatorially aspherical* if, for no $\rho \in R$, $w\rho = 1 \in F$ and if P has an irredundant part satisfying one (and hence each) of the four equivalent conditions of the above Corollary.

(CLA) P is *Cohen-Lyndon aspherical* if, for no $\rho \in R$, $w\rho = 1 \in F$ and if P has an irredundant part $\hat{P} = (X; \hat{R}, \hat{w})$ such that the normal closure $N = N(R) = N(\hat{R})$ of $\hat{w}(\hat{R})$ and $w(R)$ in F has a basis

$$B = \bigcup_{\hat{R}} \{uru^{-1} \ ; \ u \in U(r)\}$$

where, for each $r \in \hat{R}$, $U(r)$ is a full left transversal for $NC(r)$, $C(r)$ being the centraliser of r in F.

We note that the definitions given in [C-C-H] differ from the above ones (but are, of course, equivalent).

These notions are linked by the implications

$$\begin{array}{ccccc} & & SA & \Longrightarrow & \text{Aspherical} \\ & & \Downarrow & & \Downarrow \\ CLA & \Longrightarrow & DA & \Longrightarrow & CA \end{array}$$

and are studied extensively in [C-C-H]. The homotopy type of the geometric realisation of a combinatorially aspherical presentation is determined in [Hu2]. In [C-H], diagrammatically aspherical presentations are studied from a geometric point of view, and a consequence of the main result in [C-H] is that small cancellation presentations are diagrammatically aspherical. This was claimed (though in a different terminology) in the proof of Theorem III of [L4], but the proof is not correct.

## 9. Examples and an unsettled problem of J.H.C. Whitehead.

We now consider examples from §1 in the light of later sections. As explained in §6, for non-primary presentations we must distinguish between Y-sequences and $R^F$-sequences, and thus the intuitive terminology of §1 is not accurate. In the case of presentations which are irredundant and primary, the distinction between the two kinds of identity sequences is not crucial.

Example 3 of §1 was an identity between six commutators in the generators x, y, z. Suppose for precision that the presentation is irredundant with set R of relators consisting solely of the commutators [x, y], [y, z], [z, x]. Then we have the identity $R^F$-sequence over $P = (x, y, z; r_1, r_2, r_3)$

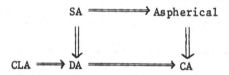

$$\underset{\sim}{p} = (r_1, r_2^y, r_3, r_4^z, r_5, r_6^x)$$

where $r_1 = [x, y]$, $r_2 = [x, z]$, $r_3 = [y, z]$, $r_4 = [y, x]$, $r_5 = [z, x]$, $r_6 = [z, y]$. So we have $r_1 = r_4^{-1}$, $r_2 = r_5^{-1}$, $r_3 = r_6^{-1}$. Note that $x$, $y$, $z$ do not belong to $N(R)$. So $\underset{\sim}{p}$ does not have the identity property and (since $P$ is irredundant and primary) we may deduce that $\underset{\sim}{p}$ represents a non-trivial element of the module $\pi$ of identities.

Example 4 of §1 was an identity among relations for the standard presentation of $\mathbb{Z}_2 \times \mathbb{Z}_2$. This presentation is not primary, so we must deal with Y-sequences, and the identity Y-sequence for this example is

$$\underset{\sim}{p} = (\rho, \tau, \sigma^{xy}, (\rho^{-1})^y, \sigma^{-1}, (\tau^{-1})^{\widetilde{\widetilde{xy}}}),$$

where $\rho, \sigma, \tau$ are elements of $R$ mapped to $r$, $s$, $t$. Note that $\widetilde{\widetilde{xy}} \notin N(R)$, so that $\underset{\sim}{p}$ does not have the identity property. Hence the corresponding element of $\bar{C}$

$$e_\rho^2(1 - \phi y) + e_\sigma^2(\phi(xy) - 1) + e_\tau^2(1 - \phi(\widetilde{\widetilde{xy}}))$$

does not belong to the root module; this may be verified directly.

In Example 5 of §1 (which illustrates an identity for the standard presentation of the symmetric group $S_3$), the relator $t$ occurs three times, and so the corresponding identity sequence will not have the identity property.

An important theorem of Lyndon [L1] is that any one-relator presentation has the identity property. In particular, if $P = (X; r)$ is a one-relator presentation, and $P$ is primary, then $P$ is aspherical. Another proof of this result is given by [D-V], who also construct other examples of aspherical spaces. In particular, they solve a problem of Papakyriakopoulos [P2] in showing that if $P$ is the presentation

$$(a, b, x_1, y_1, \ldots, x_n, y_n; [a, b] \prod_{i=1}^{n} [x_i, y_i], [a, b\tau])$$

where $\tau$ belongs to the commutator subgroup $[FX, FX]$, then $P$ is aspherical; the proof is a delicate combination of rewriting arguments and covering space techniques.

Lyndon's theorem is often called the Simple Identity Theorem. A geometric proof of a stronger theorem is given by Huebschmann in [Hu1]; it uses "pictures" (which are described in the next section).

A stronger result again is that a one-relator presentation is

CLA (§8). This result is due to Cohen-Lyndon [C-L] and is reproved in [C-C-H].

A deep geometric result of Papakyriakopoulos [P1] (the sphere theorem) implies the asphericity of certain presentations of the groups of knots and of links. Here a *link group* is the fundamental group of the complement of a tame link L in $S^3$. The link L is called *geometrically unsplittable* if there is no embedded pℓ 2-sphere $S^2$ in $S^3\backslash L$ such that each component of $S^3\backslash S^2$ contains points of L.

THEOREM [P1] : *If* L *is a link in* $S^3$, *then* $S^3\backslash L$ *is aspherical if and only if* L *is geometrically unsplittable.*

Now the link group $G = \pi_1(S^3\backslash L)$ has the *Wirtinger presentation* $P = (x_1, \ldots, x_n; r_1, \ldots, r_m)$ coming from an oriented diagram for L, with generators $x_1, \ldots, x_n$, one for each overpass, and for each crossing

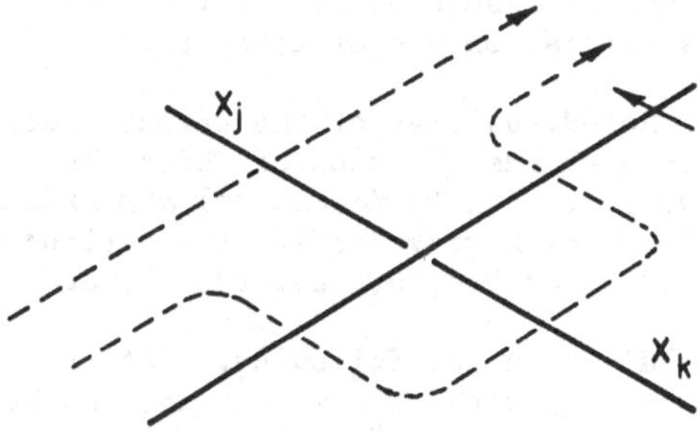


Figure 3

a relation $x_j^{-1}x_i^{-1}x_kx_i = 1$ (corresponding to a deformation between the two dotted paths shown in Fig. 3). There is always an identity among these relations, corresponding to a loop drawn right round the diagram of the link. So we have a presentation $P' = (x_1, \ldots, x_n; r_1, \ldots, r_{m-1})$ of the same group. An implication of Papakyriakopoulos' theorem is that P' is aspherical if and only if L is geometrically unsplittable. In particular if L is a knot, then P' is aspherical. For this *asphericity of knots* no purely algebraic proof is known.

In [C-C-H] it is proved, without using the sphere theorem,

that the group of any tame graph has a CLA presentation corresponding to a handle decomposition of the exterior space. An immediate consequence is the asphericity of geometrically unsplittable tame graphs.

The asphericity of knots could be easily proved if one knew that the primary identity property is *hereditary*, i.e. is inherited by subpresentations. Here if $P = (X; R)$ is a presentation, then a *subpresentation* is a group presentation $P' = (X'; R')$ for which $X' \subset X$, $R' \subset R$. Of course the group $G'$ of $P'$ may be quite different from $G$. However it is not even known if the primary identity property is inherited by $P'$ in the case when $X' = X$ and $R$ has one more element than $R'$; indeed the general finite case would follow from the special case.

The way the asphericity of knots could be deduced from this result is as follows. Add to the Wirtinger presentation of the knot the extra relation $x_1$. Geometrically, this corresponds to cutting the knot at the overpass $x_1$. The knot can then be untied. This corresponds to a combinatorial equivalence:

$$(x_1, .., x_n; r_1, .., r_{n-1}, x_1) \sim (x_1, .., x_n; x_1, .., x_n) .$$

Such equivalences preserve the module of identities, and the last presentation clearly has trivial module of identities.

The question of the hereditability of the primary identity property is equivalent to a famous question of Whitehead, raised in [Wh1]: *is every subcomplex of a 2-dimensional aspherical complex aspherical?* This seems a very difficult question: work has been done by [A, Be, B-D, C2, Co, H, Hu1, P2, S2, St].

The problem is equivalent to the following. Let $L$ be a connected 2-dimensional complex with $\pi_2 L \neq 0$. Let $K$ be formed from $L$ by attaching a set of 2-cells. Is it true that $\pi_2 K \neq 0$? (For, it is easily seen that attaching any set of 0-cells or 1-cells leaves $\pi_2$ non-zero.)

Adams [A] shows that the condition that $L$ be 2-dimensional is essential here. He sets $L = (S^1 \vee S^2) \cup_f e^3$; here $\pi_1$ is infinite cyclic generated by $z$ say, $\pi_2 L^2$ is isomorphic to the group ring of $\mathbb{Z}$ and $f$ represents the element $2 - z$ of this group ring. Let $K = L \cup_g e^2$, where $g$ represents the class $z$. Then $\pi_2 K = 0$ but $\pi_2 L$ is isomorphic with the additive group of fractions $m/2^n$, and so is non-zero.

Another possible generalisation of the question is: can $\pi_n$

of an n-complex be killed by attaching n-cells? This is easily settled if $n > 2$. For example if $K = E^3 \vee E^3$ and L is the 3-subcomplex $L = \dot{E}^3 \vee E^3$, then $\pi_3 L = \mathbb{Z}$ but $\pi_3 K = 0$. Thus the question is very much a two-dimensional one, and its difficulties are connected with our lack of understanding of crossed modules, and in particular, of free crossed modules.

In the next section, and in [St], geometric reasons are given which indicate the complexity of the problem.

The book by Lyndon and Schupp [L-S] uses the term *aspherical* in various senses. In referring to the asphericity of knots (p.162) the term is used in the sense given above. On p.157, the term is used to mean, in our terminology, that any identity $R^F$-sequence collapses to the empty sequence $\emptyset$, i.e. that the presentation is diagrammatically aspherical. However, the presentation $(x; x^2)$ of the group $\mathbb{Z}_2$ is diagrammatically aspherical but not aspherical. Further, it is not true that for any identity sequence $\underset{\sim}{y}$, "$\underset{\sim}{y}$ is equivalent to $\emptyset$" implies "$\underset{\sim}{y}$ collapses to $\emptyset$", nor is it true that "$\underset{\sim}{y}$ is Peiffer equivalent to $\emptyset$" implies "$\underset{\sim}{y}$ Peiffer collapses to $\emptyset$". Examples of this type of phenomenon were given in May, 1978, by J. Howie and by P. Stefan (private communications); later examples were given by I. Chiswell [C-C-H] and by A. Sieradski [S2] for aspherical presentations. Stefan's example is given elsewhere in this volume [St]; Howie's example is the identity sequence

$$\underset{\sim}{c} = ([x, y], [x, z]^y, [y, z], [y, x]^z, [z, x], [z, y]^x)$$

for the presentation $P = (x, y, z; [x, y], [y, z], [z, x], x, y, z)$ of the trivial group. It is easy to see that there must be a Peiffer equivalence of $\underset{\sim}{c}$ to $\emptyset$ (for example, we can check that $\underset{\sim}{c}$ has the primary identity property); a "diagram" for such an equivalence is given in Section 10. However, there is no Peiffer collapse of $\underset{\sim}{c}$ to $\emptyset$ since, by Proposition 12, exchanges on $\underset{\sim}{c}$ turn $[x, y]$ and $[y, x]^z$ to $[x, y]^u$ and $[y, x]^v$ respectively, where u has even length and v has odd length, so that this pair can never be deleted after other exchanges or collapses. Chiswell's example is the sequence

$$(y^{-2}xyx, (yx)^{-1}x^{-1}(yx), yxy^{-1}, y(y^{-2}xyx)^{-1}y^{-1})$$

for the presentation $(x, y; y^{-2}xyx, x)$ of the trivial group.

## 10. Links and pictures

Let K be a two-dimensional CW-complex. In this section we

describe some geometric representatives of elements of $\pi_2(K, K^1, a)$, and of homotopies between these representatives.

Suppose the 2-cells of $K$ are indexed by a set $R$. For each 2-cell $e^2_\rho$ choose a small disc $d_\rho$ inside it. Let $f_\rho$ be the attaching map of $e^2_\rho$, and choose a path $t_\rho$ joining a point

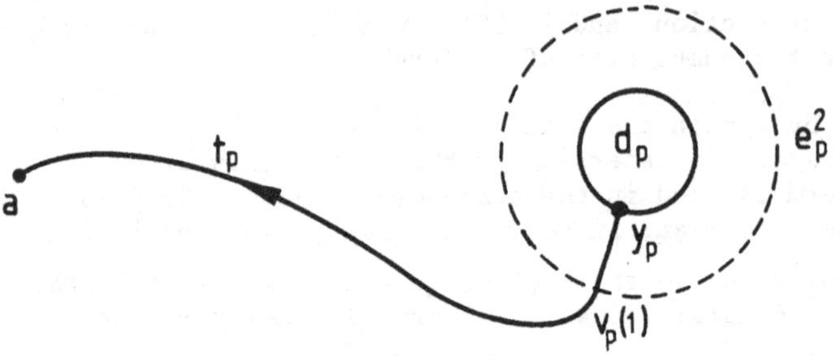


Figure 4

$y_\rho$ of $\dot{d}_\rho$ radially to $f_\rho(1)$, and then joining $f_\rho(1)$ in $K^1$ to the base point $a$ of $K$. The characteristic map for $e^2_\rho$ and the path $t_\rho$ together determine an element $a_\rho$ of $\pi_2(K, K^1, a)$, $\rho \in R$.

Let $\alpha \in \pi_2(K, K^1, a)$. Then $\alpha$ is a homotopy class of maps $k: (E^2, S^1, 1) \longrightarrow (K, K^1, a)$. It is a consequence of transversality theory (for more details of which, see [B-R-S], Ch. VII) that $\alpha$ contains a representative $k$ such that for each 2-cell $e^2_\rho$ of $K$, $k^{-1}(d_\rho)$ (where $d_\rho$ is as above) is a finite disjoint union of discs $\delta_{\rho,1}, \delta_{\rho,2}, \ldots$ each of which is mapped by $k$ homeomorphically to $d_\rho$. (Since $E^2$ is compact, $k^{-1}(e^2_\rho)$ is non-empty for only finitely many 2-cells $e^2_\rho$.) For each $i$, let $x_{\rho,i}$ be the unique point of $\delta_{\rho,i}$ such that $k(x_{\rho,i}) = y_\rho$, and in $E^2$ join each $x_{\rho,i}$ to $1$ by a path $s_{\rho,i}$ so that the various $s_{\rho,i}$ meet each other only at their final point $1$, and meet the union of the discs $\delta$ only at their initial point. Now relabel the $\delta_{\rho,i}$, $s_{\rho,i}$ as $\delta_j$, $s_j$, taking the paths in order around $1$, and let $e^2_{\rho_j}$ be the cell of $K$ containing $k(\delta_j)$. The path $-t_{\rho_j} + k(s_j)$ misses the centres of all $e^2_\rho$; it therefore can be deformed radially off each $e^2_\rho$ into $K^1$, so determining a class $u_j$ in $\pi_1(K^1, a)$.

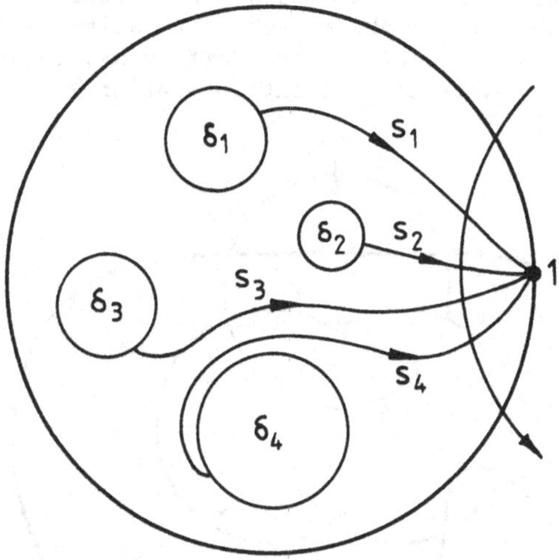


Figure 5

Let $\varepsilon_j$ be $\pm 1$ according as $k$ maps $\delta_j$ in an orientation preserving or reversing way to $e^2_{\rho_j}$ (the orientations of the 2-cells of $K$ are determined by the characteristic maps and an orientation of the standard 2-cell). Then we have an element

$$c = (\rho_1^{\varepsilon_1})^{u_1} \ldots (\rho_n^{\varepsilon_n})^{u_n}$$

of the free crossed $\pi_1(K^1, a)$-module $C$.

Let $\phi\colon C \longrightarrow \pi_2(K, K^1, a)$ be the map of crossed $\pi_1(K^1, a)$-modules such that $\phi(\rho) = a_\rho$, $\rho \in R$ (cf. §5). Then if $\alpha, k, c$ are as above, the homotopy addition lemma in dimension 2 implies that $\phi(c) = \alpha$. This explains why $\phi$ is surjective, a fact used in our proof of the special case of Theorem 10.

Suppose now that $F\colon (E^2, S^1, 1) \times I \longrightarrow (K, K^1, a)$ is a homotopy such that $F_0$, $F_1$ satisfy the properties of the map $k$ above. Then $F$ may be deformed rel $E^2 \times \dot{I}$ to a map $G$ so that for all $\rho \in R$, $G^{-1}(d_\rho)$ is a disjoint union of solid tubes $\delta \times I$ and solid tori $\delta \times S^1$ (where $\delta$ is a 2-disc), so that $G$ restricted to one of these is projection to $\delta$ followed by a homeomorphism to $d_\rho$. As pointed out in [S2], the union of these $\delta \times I$ and $\delta \times S^1$ for all $e^2_\rho$ is a framed link in $E^2 \times I$. The use of this idea by Whitehead in [Wh1] (cf. [Br1]) suggested to Stefan and to Sieradski [St, S2] a geometric interpretation of Peiffer moves. We illustrate the method using slightly different conventions to those of [S2, St] . (R. Peiffer has informed us that Reidemeister was aware of this interpretation.)

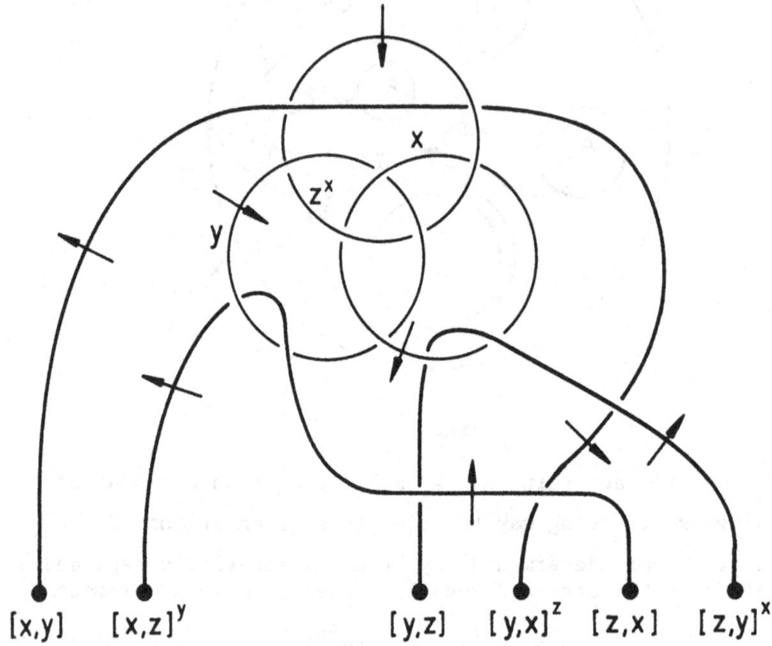


Figure 6

EXAMPLE: The above is a labelled, oriented link diagram with six "feet". Not all the labels have been inserted, because the remaining ones can all be deduced from the rules :

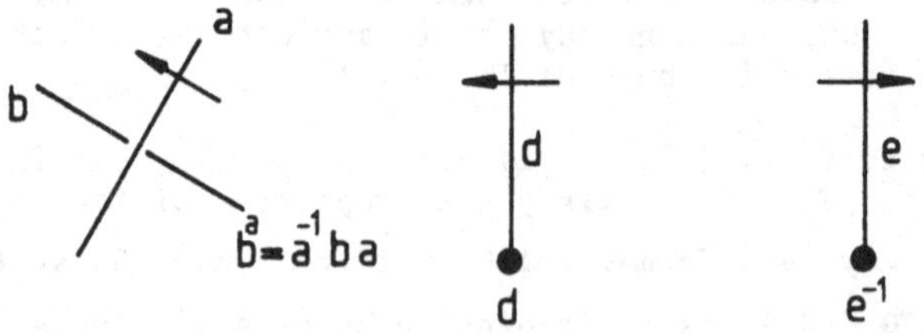


(i) (ii) (iii)

The labels are to be conjugates of relators, considered as elements of the free group FX on the generators. With this convention, the above diagram is consistent, in that no overpass has distinct labels in FX ; the checking of this is left as an exercise to the reader.

The diagram determines an equivalence of $\underset{\sim}{c}$ to $\emptyset$ for the presentation $P = (x, y, z; [x, y], [y, z], [z, x], x, y, z)$ of the trivial group, where $\underset{\sim}{c}$ is the identity sequence given at the base of the diagram. Successive identity sequences in this equivalence are found by horizontal cuts of the diagram, at different heights and in general positions; the identity sequence corresponding to a cut is read off by rules (ii) and (iii). (The diagram is a simplification by R. Brown of a diagram of an equivalence of $\underset{\sim}{c}$ to $\emptyset$ with about 100 crossings, drawn by J. Howie.)

The reader will note the appearance of the Borromean rings in Fig. 6. They appear for a similar reason in [F-T]. The reader is invited to try and find an equivalence of $\underset{\sim}{c}$ to $\emptyset$ for which the corresponding diagram has the inserted $x, y, z$ less subtly linked.

The above diagram is a partial representation of a null-homotopy $G: (E^2, S^1, 1) \times I \longrightarrow (K, K^1, a)$. We have drawn only the centre lines $0 \times I$, $0 \times S^1$ (where $0$ is the centre of $\delta$) of the components of the framed link, and the link itself takes account only of the 2-cells of $K$ and not the 1-cells. Rourke in [Rou] has developed the above use of what is essentially transversality to give a more detailed description of maps $k: (E^2, S^1, 1) \longrightarrow (K, K^1, a)$ in terms of "pictures". We explain the idea for those 2-complexes that are geometric realisations of presentations.

Consider the presentation $(X; R) = (x, y; r, s, t)$ of the trivial group, where $r = x^2y$, $s = y^{-1}x$, $t = x$. Let $K = K(X; R)$ be its geometric realisation, with vertex $a$. Here is an example of a "picture" of a particular map $k: (E^2, S^1, 1) \longrightarrow (K, K^1, a)$ with $k(S^1) = \{a\}$ :

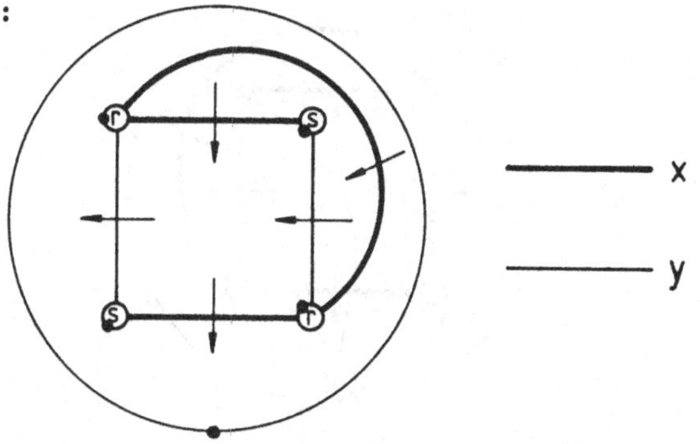


Figure 7

The constituents of such a picture over a presentation (X; R, w) are:

(i) a disc D with boundary $\partial D$ ;

(ii) disjoint discs $\Delta, \Delta_1, \Delta_2, \ldots, \Delta_n$ inside D , each labelled by an element of R or, in the case of an irredundant presentation, by a relator;

(iii) disjoint edges $e, e_1, \ldots, e_m$ inside D and outside the $\Delta$'s ; each edge is either a circle, or joins the boundaries of two of the discs in (i) or (ii) ; each edge has a normal orientation, indicated by a short arrow meeting the edge transversally, and is also labelled by a 1-cell of $K^1$ (identified with a generator of $\pi_1(K^1)$ ) ;

(iv) base points, indicated by a dot, on the boundaries of D and of each of the $\Delta$'s , but not lying on any edge.

The one further condition imposed is that starting from a base point of some $\Delta$ and reading the oriented edges round $\Delta$ in an anticlockwise direction should give the relator, or its inverse, labelling that disc.

A picture is called *spherical* if it has no edges meeting $\partial D$ ; so the picture of Figure 7 is spherical.

Any picture over a presentation (X; R, w) determines some Y-sequences over (X; R, w) and if the picture is spherical these Y-sequences are identity sequences. We illustrate this process first for the picture of Figure 7.

For each $\Delta$ draw a line from the base point of $\Delta$ to that of D so that these lines cross the edges transversally and meet only at the base point of D . This gives for example, the next figure:

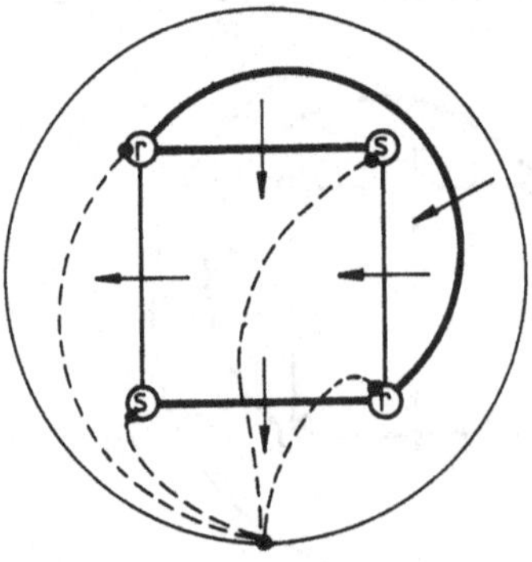


Figure 8

To each dotted line $\alpha$ we can associate a symbol $(\rho^u)^\varepsilon$ . Here $\rho$ is the label of the disc from which $\alpha$ starts. This disc also has a sign $\varepsilon$ which is $+1$ or $-1$ according as reading anticlockwise round the disc, starting at the base point, gives $w\rho$ or $(w\rho)^{-1}$ . The element $u$ of $F$ is the product of the labels of the edges that $\alpha$ crosses, taken in order from the initial disc of $\alpha$ , and with a sign $+1$ or $-1$ according as $\alpha$ crosses the edge in a positive or negative normal direction.

The dotted edges have an order, obtained by reading them anticlockwise round the base point of $D$ . From this order, and the associated symbols, we obtain a Y-sequence. This gives for Fig. 8 the identity sequence:

$$\underset{\sim}{y} = (r^x, s^x, s^{-1}, r^{-1}) .$$

The Peiffer transformations now have the following interpretations. A Peiffer insertion, e.g.

$$\underset{\sim}{y} = (r^x, s^x, x^{-1}, r^{-1}) \longmapsto (t^{-1}, t, r^x, s^x, s^{-1}, r^{-1}) = \underset{\sim}{z}$$

corresponds to introducing in Fig. 8 a new component with only two discs as indicated in Fig. 9 (where we now omit to draw $\partial D$) :

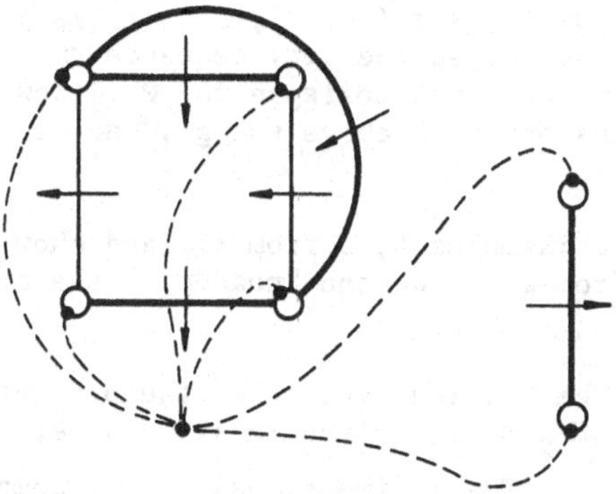

Figure 9

Conversely, we can carry out a deletion on Fig. 9 . (A general description of insertions and deletions is given later.)

An elementary Peiffer exchange, e.g.

$$(t^{-1}, t, r^x, s^x, s^{-1}, r^{-1}) \longmapsto (t^{-1}, r, t, s^x, s^{-1}, r^{-1})$$

corresponds to rechoosing two successive lines from base points of discs $\Delta$ to the base point of $D$. In our example, the result is indicated in Fig. 10.

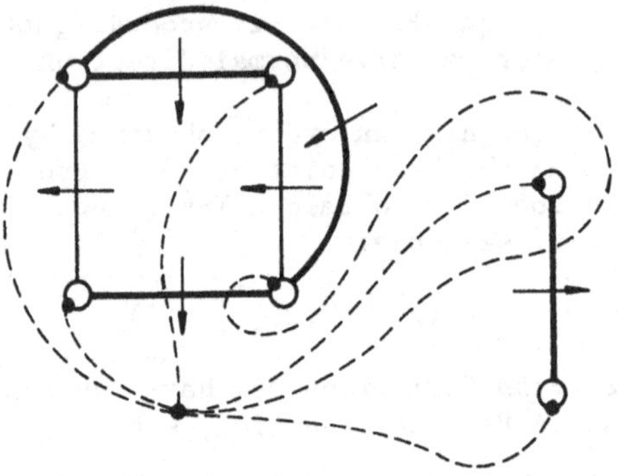

Figure 10

Hence a Peiffer exchange corresponds to rechoosing the paths from the base points of the $\Delta$'s to that of $D$.

The identity sequence $y = (r^x, s^x, s^{-1}, r^{-1})$ given above is P. Stefan's example [St] of an identity sequence that is Peiffer equivalent to $\emptyset$, but does not collapse to $\emptyset$. However, by an insertion $y$ is transformed as above to $\underset{\approx}{z}$, and $\underset{\approx}{z}$ does collapse to $\emptyset$ [St]$^{\sim}$.

We now return to Examples 4, 5 from §1, and show further how to obtain pictures from which we can "read off" the corresponding identity sequence.

EXAMPLE 4. Let $K$ be the realisation of the presentation $(x, y; r, s, t)$ of $\mathbb{Z}_2 \times \mathbb{Z}_2$ given in §1, and let $\tilde{K}$ be the universal cover of $K$. The 1-dimensional Cayley complex given in Fig. 1 is the 1-skeleton $\tilde{K}^1$ of $\tilde{K}$, and the labels of the edges determine the covering map $\tilde{K}^1 \longrightarrow K^1$. Using Fig. 1 we can regard $\tilde{K}^1$ as contained in $S^2$ (taken as a disc with boundary identified to a point), and the map $\tilde{K}^1 \longrightarrow K$ extends to a map $f: S^2 \longrightarrow K^2$ in which the regions in which $\tilde{K}^1$ divides $S^2$ are mapped to the labelled cells, the outside of $\tilde{K}^1$ being mapped to $t^{-1}$. This map corresponds to a spherical picture which arises in essence as dual to the Cayley complex. This picture is given by the thick lines in Fig. 11; the thin lines give the Cayley diagram, the dots denote base points, and the dotted lines are used to determine a Y-sequence from the picture, as described earlier. The resulting Y-sequence is precisely the identity sequence given in §1 .

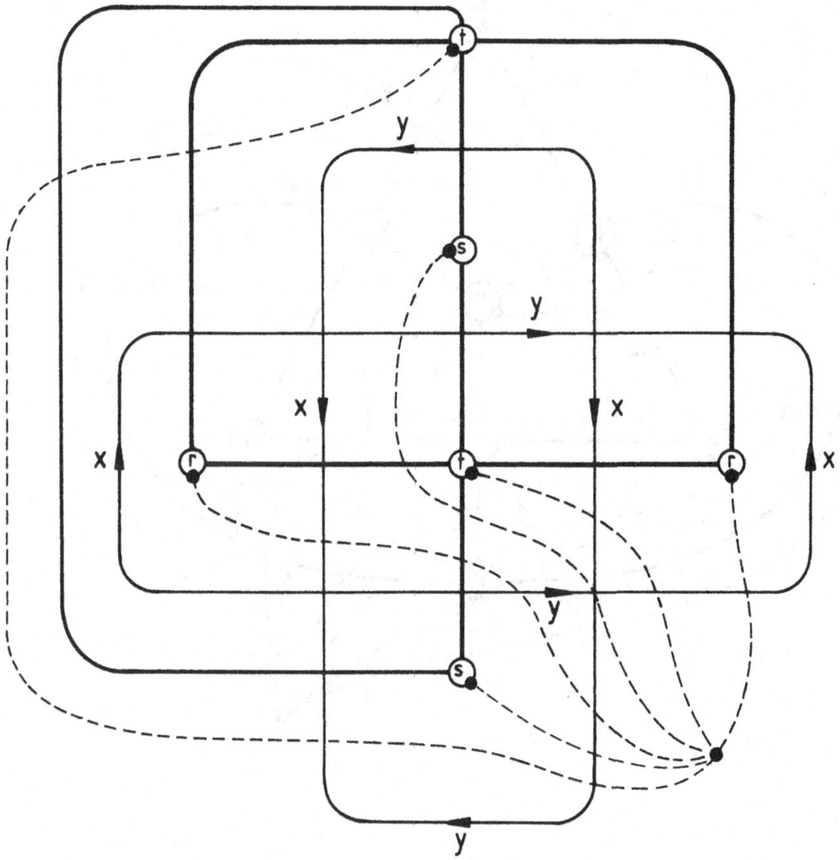


Figure 11

EXAMPLE 5. In this example we give only the picture corresponding to the Cayley diagram, together with the dotted curves which determine the corresponding identity sequence.

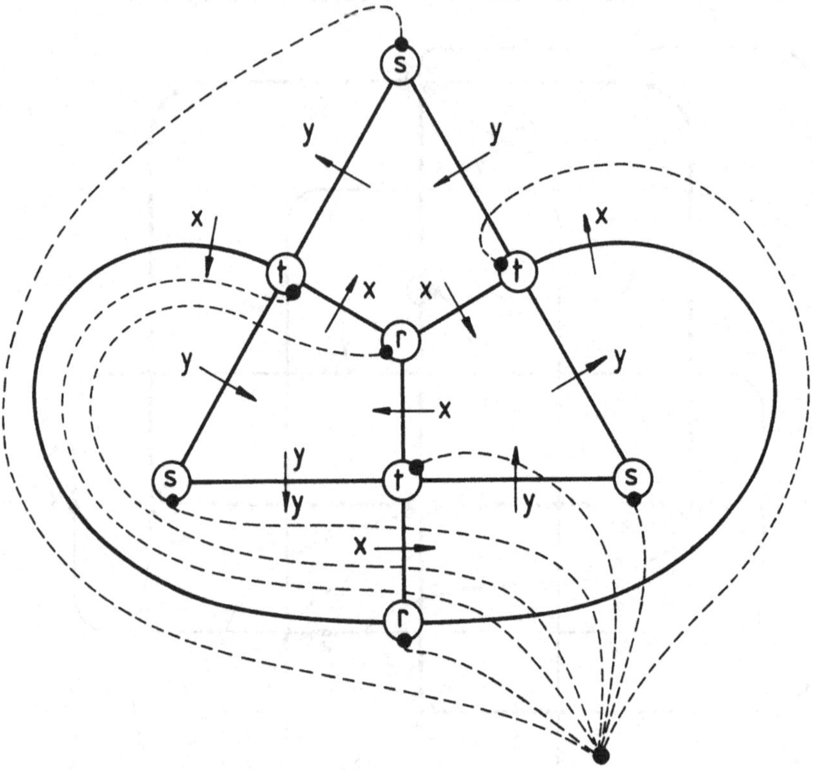


Figure 12

We now sketch the ideas of "deformations" of pictures, which correspond to homotopies. These are:

(D1) Removal of edges which are loops enclosing no discs or edges e.g.

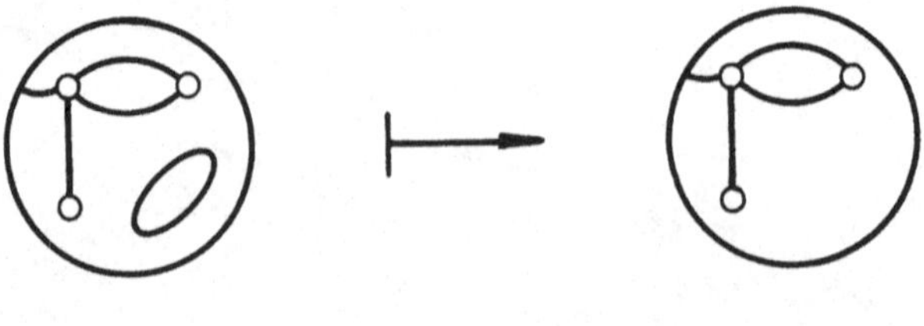

(D1') Insertion of such a 'floating circle'.

(D2) Bridge moves:

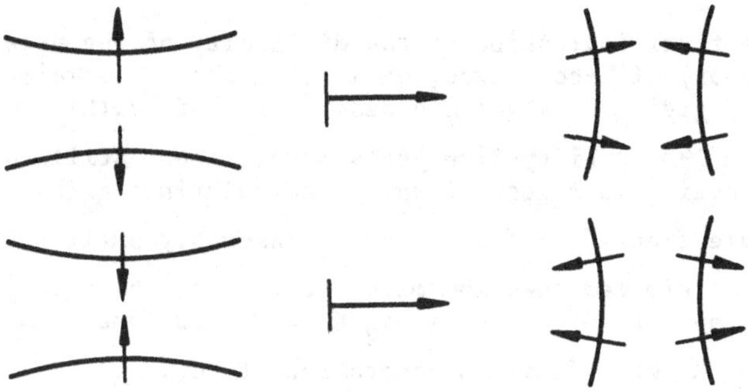

(D3) Removal of a 'floating component' enclosing no other discs or edges, such as

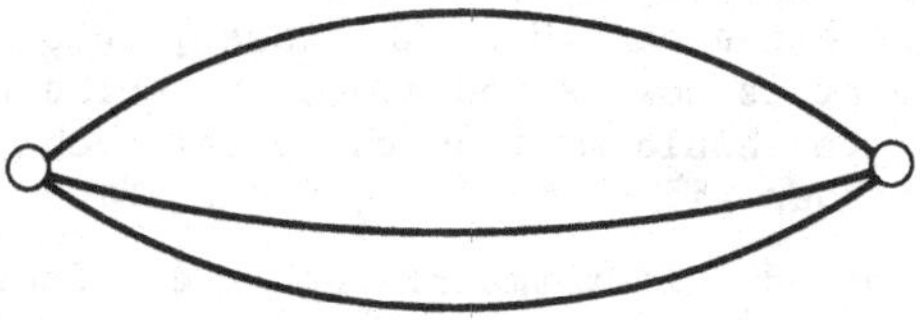

(D3') Reverse of (D3) .

The operation (D3) corresponds to a Peiffer deletion, but may be applicable only after a sequence of bridge moves.

These pictures and deformations are exploited by Rourke [Rou] and Huebschmann [Hu1] .

In particular, [Hu1] uses them to give examples of aspherical 2-complexes for which every subcomplex is aspherical. No such families were known before.

The hereditability of the primary identity property, which was discussed in the last section, can be expressed in terms of pictures and deformations as follows. Let $(X'; R')$ be a sub-presentation of the aspherical presentation $(X; R)$ . Let $P$ be a picture for an identity sequence $\underset{\sim}{y}$ for $(X'; R')$ . Since $(X; R)$ is aspherical, there is a sequence of deformations $P \longmapsto P_1 \longmapsto \dots \longmapsto P_n = \emptyset$ , which may involve moves (D1') or (D3') using edges labelled by elements of $X \backslash X'$ , or discs labelled by elements of $R \backslash R'$ . The hereditability of the primary identity property would imply that there is also a sequence of deformations $P \longmapsto P'_1 \longmapsto \ldots \longmapsto P'_m = \emptyset$ involving labels only from $X'$ and $R'$.

As one final indication of the difficulty of the area of the homotopy theory of 2-complexes, we mention that Reidemeister's paper [Re1], giving an algebraic description of $\pi_2(K)$, was published in 1934. Fifty-five years later, there still does not seem to be available a general way of calculating $\pi_2(K)$ as a $\pi_1(K)$ module even if $\pi_1(K)$ is some reasonably small finite group $G$. A simpler question might be to ask: which complex representations of $G$ arise as $\pi_2(K) \otimes \mathbb{C}$ for some geometric realisation $K$ of a finite presentation of $G$?

The history of the methods described in this article is complex, and is to some extent shown by the references given throughout. In effect, the use of the chains of the universal cover is due to Reidemeister [Re1]. His work and that of his students developed in Eilenberg-MacLane's work on complexes with operators, and, in the hands of J.H.C. Whitehead, into simple homotopy theory and what is now termed algebraic K-theory. Coming to the present field, we should mention the paper [Sch], which contains the result of our §5 that $\operatorname{Im} d_2 \cong \underline{N}$. Note that the description of $d_2$ in terms of the free differential calculus is given in [Fo]. An equivalent, and earlier, formulation is due to Whitehead, in that Theorem 8 of [Wh3] gives a clear and complete description of the relationship between a free crossed module over a free group and the associated chain complex.

There are two useful generalisations of the embedding of the relation module $\overline{N}$ into the free module $\underset{X}{\oplus} \mathbb{Z}G$. One of these, described lucidly in [Cr], assigns to any exact sequence of groups $1 \longrightarrow N \longrightarrow \Gamma \xrightarrow{\phi} G \longrightarrow 1$ an exact sequence of G-modules $0 \longrightarrow \overline{N} \longrightarrow D_\phi \longrightarrow IG \longrightarrow 0$. (This is in fact Satz 15 of [Sch].) Another is the Magnus embedding of a free group into a matrix group [Ma]; this is applied in [B1, D2] and an account of the generalisation to the case of a homomorphism $\phi: \Gamma \longrightarrow G$, is given in [Bi] §3.2, where further references may be found.

Acknowledgements

We would like to thank I.M. Chiswell, M.M. Cohen, D.J. Collins, R. Fenn, P.J. Higgins, J. Howie, R. Peiffer and C.T.C. Wall for helpful comments on a draft of this paper, and in particular thank P.J. Higgins for his formulation of the proof of Proposition 3.

SUMMARY OF NOTATIONS.

$P = (X; R, w)$ is a presentation of a group $G$. The sequence

$$1 \longrightarrow E \longrightarrow H \xrightarrow{\theta} F \xrightarrow{\phi} G \longrightarrow 1$$

is exact where $H = F(X \times R)$, $F = F(X)$.
The Peiffer group $P$ is normal in $H$; $N = \operatorname{Im}\theta$. There is a diagram of short exact sequences

$$\begin{array}{ccccccc}
 & & 1 & & 1 & & \\
 & & \downarrow & & \downarrow & & \\
 & & P & = & P & & \\
 & & \downarrow & & \downarrow & & \\
1 & \longrightarrow & E & \longrightarrow & H & \longrightarrow & N \longrightarrow 1 \\
 & & \downarrow & & \downarrow & & \\
1 & \longrightarrow & \pi & \longrightarrow & C & \longrightarrow & N \longrightarrow 1 \\
 & & \downarrow & & \downarrow & & \\
 & & 1 & & 1 & &
\end{array}$$

The free crossed $F$-module $C \xrightarrow{\partial} F$ is isomorphic to

$$\pi_2(K, K^1) \longrightarrow \pi_1(K^1) \quad \text{where} \quad K = K(P).$$

There is a diagram of $G$-modules, with exact rows:

$$\begin{array}{ccccccc}
0 \longrightarrow \pi \longrightarrow & \bar{C} & \longrightarrow & \bar{N} & \longrightarrow & 0 & \\
 & \downarrow \cong & & \downarrow i & & & i \text{ injective} \\
 & \oplus\, \mathbb{Z}G & \xrightarrow{d_2} & \oplus\, \mathbb{Z}G & \xrightarrow{d_1} & \mathbb{Z}G & \\
 & \downarrow \cong & & \downarrow \cong & & \downarrow \cong & \\
 & C_2(K) & \longrightarrow & C_1(K) & \longrightarrow & C_0(K) &
\end{array}$$

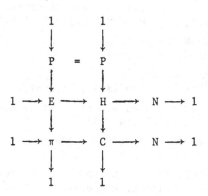

# REFERENCES


[A] J.F. ADAMS, 'A new proof of a theorem of W.H. Cockcroft', *J. London Math. Soc.* 30 (1955), 482-488.

[Ba] H. BASS, 'Traces and Euler characteristics', in *Homological Group Theory,* Proc. Durham Symp. (1977) *Ed.* C.T.C. WALL, London Math. Soc. Lecture Note Series No. 36 (1979), 1-26, Cambridge Univ. Press.

[Be1] W.H. BECKMANN, 'A certain class of non-aspherical 2-complexes', *J. Pure Appl. Alg.* 16 (1980), 243-244.

[Be2] W.H. BECKMANN, *Completely aspherical 2-complexes,* Ph.D. Thesis, Cornell U. (1980).

[Bi] J.S. BIRMAN, *Braids, links and mapping class groups,* Annals of Math. Studies(1974), Princeton Univ. Press.

[Bl] N. BLACKBURN, 'Note on a theorem of Magnus', *J. Austral. Math. Soc.* 10 (1960), 469-474.

[B-D] J. BRANDENBURG and M.N. DYER, 'On the aspherical Whitehead conjecture', Comm. Math. Helv. 56 (1981),431-446

[Be-Du] P.H. BERRIDGE and M.J. DUNWOODY, 'Non-free projective modules for torsion-free groups', *J. London Math. Soc.* (2), 19 (1979), 433-436.

[Br1] R. BROWN, 'On the second relative homotopy group of an adjunction space: an exposition of a theorem of J.H.C. Whitehead', *J. London Math. Soc.* (2), 22 (1980), 146-152.

[Br2] R. BROWN, 'Higher dimensional group theory', (this volume).

[B-H1] R. BROWN and P.J. HIGGINS, 'On the connection between the second relative homotopy groups of some related spaces', *Proc. London Math. Soc.* (3), 36 (1978), 193-212.

[B-H2] R. BROWN and P.J. HIGGINS, 'Colimit theorems for relative homotopy groups', *J. Pure Appl. Alg.*, 22 (1981), 11-41.

[B-R-S] S. BUONCRISTANO, C.P. ROURKE and B.J. SANDERSON, *A geometric approach to homology theory,* London Math. Soc. Lecture Note Series No. 18 (1976), Cambridge Univ. Press.

[C-C-H] I.M. CHISWELL, D.J. COLLINS and J. HUEBSCHMANN, 'Aspherical group presentations', *Math.Z.* 178 (1981),1-36.

[C1] W.H. COCKCROFT, 'Note on a theorem due to J.H.C. Whitehead', *Quart. J. Math.* 2 (1951), 159-160.

[C2] W.H. COCKCROFT, 'On two-dimensional aspherical complexes', *Proc. London Math. Soc.* (3), 4 (1954), 375-384.

[Co] J.M. COHEN, 'Aspherical 2-complexes', *J. Pure Appl. Alg.* 12 (1978), 101-110.

[Cr] R.H. CROWELL, 'The derived module of a homomorphism', *Advances Math.* 6 (1971), 210-238.

[C-F] R.H. CROWELL and R.H. FOX, *Introduction to knot theory*, Ginn and Co., (1953).

[C-H] D.J. COLLINS and J. HUEBSCHMANN, 'Spherical diagrams and identities among relations', (preprint, 1981).

[C-L] D.E. COHEN and R.C. LYNDON, 'Free bases for normal subgroups of free groups', *Trans. Amer. Math. Soc.* 108 (1963), 528-537.

[D1] M.J. DUNWOODY, 'On relation groups', *Math. Z.* 81 (1963), 180-186.

[D2] M.J. DUNWOODY, 'The Magnus embedding', *J. London Math. Soc.* 44 (1969), 115-117.

[D3] M.J. DUNWOODY, 'Relation modules', *Bull. London Math. Soc.* 4 (1972), 151-155.

[D4] M.J. DUNWOODY, 'The homotopy type of a two-dimensional complex', *Bull. London Math. Soc.* 8 (1976), 282-285.

[D5] M.J. DUNWOODY, 'Answer to a conjecture of J.M. Cohen', *J. Pure Appl. Alg.* 16 (1980), 249.

[D-V] E. DYER and A.T. VASQUEZ, 'Some small aspherical spaces', *J. Austral. Math. Soc.* 16 (1973), 332-352.

[F-T] R. FENN and P. TAYLOR, 'Introducing doodles', in *Topology of low-dimensional manifolds*, Proceedings of the second Sussex Conference, 1977 *Ed.* R. FENN, Springer Lecture Notes in Math. 722 (1979), 37-43.

[Fo] R.H. FOX, 'Free differential calculus I, Derivation in the free group ring', *Annals of Math.* 57 (1953), 547-560.

[G1] K.W. GRUENBERG, *Cohomological topics in group theory*, Lecture Notes in Math. 143, Springer-Verlag, Berlin-Heidelberg, New York (1970).

[G2] K.W. GRUENBERG, 'Relation modules of finite groups', CBMS No. 25, *Amer. Math. Soc.* Providence, R.I. (1976).

[G3] K.W. GRUENBERG, 'Free abelianised extensions of finite groups', in *Homological group theory*, *Ed.* C.T.C. WALL, Proc. Durham Symp. (1977), London Math. Soc. Lecture Note Series 36 (1979), 71-104, Cambridge Univ. Press.

[G-R] M.A. GUTIERREZ and J.G. RATCLIFFE, 'On the second homotopy group', *Quart. J. Math.* (2) 32 (1981), 45-56.

[Hi] P.J. HILTON, *An introduction to homotopy theory*, Cambridge Univ. Press (1953).

[Hi-St] P.J. HILTON and U. STAMMBACH, *A course in homological algebra*, Graduate texts in Mathematics 4, Springer,

Berlin (1970).

[H] J. HOWIE, 'Aspherical and acyclic 2-complexes', *J. London Math. Soc.* (2), 20 (1979), 549-558.

[Hu1] J. HUEBSCHMANN, 'Aspherical 2-complexes and an unsettled problem of J.H.C. Whitehead', Math. Ann. 258(1981), 17-38.

[Hu2] J. HUEBSCHMANN, 'The homotopy type of a combinatorially aspherical presentation', *Math. Z.* 173 (1980), 163-169.

[Hu3] J. HUEBSCHMANN, 'Cohomology theory of aspherical groups and of small cancellation groups', *J. Pure Appl. Algebra* 14 (1979), 137-143.

[J] D.L. JOHNSON, *Presentations of groups*, London Math. Soc. Lecture Note Series 22 (1976), Cambridge Univ. Press.

[K] R. KIRBY, 'Problems in low dimensional manifold theory', Proceedings of Symposia in Pure Mathematics, *American Math. Soc.*, 32 (1978), 273-312.

[Le] J. LEWIN, 'Projective modules over group-algebras of one-relator groups', *Abstracts Amer. Math. Soc.* 1 (1980), 617.

[Lo1] S.J. LOMONACO, Jr., 'The second homotopy group of a spun knot', *Topology* 8 (1969), 95-98.

[Lo2] S.J. LOMONACO, Jr, 'Homology of group systems with applications to low-dimensional topology', *Bull. Amer. Math. Soc.* (N.S.) 3 (1980), 1049-1052.

[Lo3] S.J. LOMONACO, Jr., 'The homotopy groups of knots I: How to compute the algebraic 3-type', *Pacific J. Math.*

[L1] R.C. LYNDON, 'Cohomology theory of groups with a single defining relation', *Annals of Math.* 52 (1950), 650-665.

[L2] R.C. LYNDON, 'Dependence and independence in free groups', *J. Reine Angew. Math.* 210 (1962), 148-174.

[L3] R.C. LYNDON, 'On the Freiheitssatz', *J. London Math. Soc.* (2) 5 (1972), 95-101.

[L4] R.C. LYNDON, 'On Dehn's algorithm', *Math. Ann.* 166 (1966), 208-228.

[L-S] R.C. LYNDON and P.E. SCHUPP, *Combinatorial group theory*, Ergebnisse der Mathematik und ihrer Grenzgebiete, Vol.89 Springer : Berlin-Heidelberg - New York, 1977.

[M] S. MACLANE, *Homology*, Die Grundlehren der Mathematischen Wissenschaften , Vol.114, Springer : Berlin-Heidelberg - New York, 1975.

[Ma] W. MAGNUS, 'On a theorem of Marshall Hall', *Annals of Math.* 40 (1939), 764-768.

[Me1] W. METZLER, 'Über den Homotopietyp zweidimensionaler CW-Komplexe und Elementartransformationen bei Darstellungen von Gruppen durch Erzeugende und definierende Relationen', *J. Reine und Angew. Math.* 285 (1976), 7-23.

[Me2] W. METZLER, 'Äquivalenzklassen von Gruppenbeschreibungen, Identitäten und einfacher Homotopietyp in niederen Dimensionen', in *Homological Group Theory*, Proc. Durham Symp. (1977), *Ed.* C.T.C. WALL, London Math. Soc. Lecture Note Series No. 36, (1979), 291-326, Cambridge Univ. Press.

[P1] C.D. PAPAKYRIAKOPOULOS, 'On Dehn's lemma and the asphericity of knots', *Ann. of Math.* 66 (1957), 1-26.

[P2] C.D. PAPAKYRIAKOPOULOS, 'Attaching 2-dimensional cells to a complex', *Ann. of Math.* 78 (1963), 205-222.

[Pe] R. PEIFFER, 'Über Identitäten zwischen Relationen', *Math. Ann.* 121 (1949), 67-99.

[R] J.G. RATCLIFFE, 'Free and projective crossed modules', *J. London Math. Soc.* (2), 22 (1980), 66-74.

[Re1] K. REIDEMEISTER, 'Homotopiegruppen von Komplexen', *Abh. Math. Sem. Univ. Hamburg* 10 (1934), 211-215.

[Re2] K. REIDEMEISTER, 'Über Identitäten von Relationen', *Abh. Math. Sem. Univ. Hamburg* 16 (1949), 114-118.

[Re3] K. REIDEMEISTER, 'Complexes and homotopy chains', *Bull. Amer. Math. Soc.* 56 (1950), 297-307.

[Rol] D. ROLFSEN, *Knots and links*, Publish or Perish, Berkeley, (1976).

[Rou] C. ROURKE, 'Presentations and the trivial group', in *Topology of Low Dimensional Manifolds*, Proceedings of the Second Sussex Conference, 1977, *Ed.* R. FENN, Springer Lecture Notes in Math. No. 727 (1979), 134-143.

[Sch] H.G. SCHUMANN, 'Über Moduln und Gruppenbilder', *Math. Ann.* 114 (1937), 385-413.

[S1] A.J. SIERADSKI, 'Combinatorial isomorphisms and combinatorial homotopy equivalences', *J. Pure Appl. Alg.* 7 (1976), 59-95.

[S2] A.J. SIERADSKI, 'Framed links for Peiffer identities', *Math. Z.* 175 (1980), 125-137.

[S-D] A.J. SIERADSKI and M.N. DYER, 'Distinguishing arithmetic for certain stably isomorphic modules', *J. Pure Appl. Algebra* 15 (1979), 199-217.

[St] P. STEFAN, 'On Peiffer transformations, link diagrams, and a question of J.H.C. Whitehead', (this volume).

[T] H.F. TROTTER, 'Homology of group systems with applications to knot theory', *Ann. of Math.* 76 (1962), 464-498.

[Wald] F. WALDHAUSEN, ' Algebraic K-theory of generalised free products, Parts I, II ', *Ann. of Math.* 108 (1978), 135-256.

[Wa1] C.T.C. WALL, 'Formal deformations', *Proc. London Math. Soc.* (3) 16 (1966), 342-352.

[Wa2] C.T.C. WALL, 'List of problems', in *Homological Group Theory*, Proc. Durham Symp. (1977), *Ed.* C.T.C. WALL, London Math. Soc. Lecture Note Series No. 36 (1979), 369-394.

[We] P. WEBB, 'The minimal relation module of a finite abelian group', *J. Pure Appl. Algebra* 21 (1981), 205-232.

[W] G.W. WHITEHEAD, *Elements of homotopy theory*, Graduate texts in Math. Vol. 61, Springer Verlag, (1978).

[Wh1] J.H.C. WHITEHEAD, 'On adding relations to homotopy groups', *Annals of Math.* 42 (1941), 409-428.

[Wh2] J.H.C. WHITEHEAD, 'Note on the previous paper', *Annals of Math.* 47 (1946), 806-810.

[Wh3] J.H.C. WHITEHEAD, 'Combinatorial homotopy II ', *Bull. Amer. Math. Soc.* 55 (1949), 453-496.

[Wr] P. WRIGHT, 'Group presentations and formal deformations', *Trans. Amer. Math. Soc.* 208 (1975), 161-169.

*Note added in proof:* K. Igusa in a preprint on 'The generalised Grassman invariant' also has described pictures like those of §10 and has used these for giving an explicit description of the exotic element in $K_3(\mathbb{Z}) = H_3(St(\mathbb{Z})) = \mathbb{Z}_{43}$ .

# On Peiffer transformations, link diagrams and a question of J. H. C. Whitehead

P. STEFAN

Introduction by Ronald Brown.

Whitehead's famous question: is every subcomplex of a 2-dimensional aspherical complex aspherical? was discussed at Bangor in March, 1978, during a visit of Johannes Huebschmann. He suggested a possible approach to this question, but doubts were raised about this in May, 1978, in a letter to me from Eldon Dyer. Peter Stefan then pinpointed precisely the failure of the proposed method, by finding an example of an identity sequence which was equivalent to Peiffer transformations to the empty sequence $\emptyset$ but which did not collapse to $\emptyset$. He sent the example to several people, and Roger Fenn in replying explained the method of pictures. Peter wrote to Roger on 17 May, 1978, and circulated this letter. Peter died in a mountaineering accident on June 18.

This note contains Peter's example, and the major part of his letter, omitting some irrelevant matters or outdated points. The article "Identities among relations" by R. Brown and J. Huebschmann, in this volume, and referred to here as [Br-Hu], is intended to give the background required for understanding this note, and so some of Peter's notations, conventions and diagrams have been changed to make the two articles consistent. Other changes are few and minor. References here other than to [Br-Hu] are to the bibliography of that article.

## 1. A Peiffer trivial identity sequence which does not collapse.

Consider the presentation $(x, y; r, s, t)$ of the trivial group in which $r = x^2y$, $s = y^{-1}x$, $t = x$. Consider the identity sequence

$$\underset{\sim}{u} = (r^x, s^x, s^{-1}, r^{-1}) .$$

CLAIM. There is an equivalence of $\underset{\sim}{u}$ to $\emptyset$, but $\underset{\sim}{u}$ does not collapse to $\emptyset$.

*Proof.* Note first that $rs = x^3$ which commutes with $x$ . So we have Peiffer transformations (where $\tilde{a}$ denotes $a^{-1}$)

$$(r^x, s^x, s^{-1}, r^{-1}) \mapsto (t^{-1}, t, r^x, s^x, s^{-1}, r^{-1}) \mapsto (t^{-1}, r, t, s^x, s^{-1}, r^{-1})$$

$$\mapsto (t^{-1}, r, s, t, s^{-1}, r^{-1}) \mapsto (t^{-1}, r, s, s^{-1}, t^{\tilde{s}}, r^{-1})$$

$$\mapsto (t^{-1}, r, s, s^{-1}, r^{-1}, t^{\tilde{s}\tilde{r}}) \mapsto (t^{-1}, r, r^{-1}, t^{\tilde{s}\tilde{r}}) \mapsto (t^{-1}, t^{\tilde{s}\tilde{r}}) .$$

However $\tilde{s}\tilde{r} = x^{-3}$ . So this last identity sequence is equivalent to $\emptyset$ , and we have an equivalence $\underset{\sim}{u} \wedge \emptyset$ .

To prove that $\underset{\sim}{u}$ does not collapse to $\emptyset$ , note that the presentation is irredundant and primary. In order to obtain a collapse, we must perform Peiffer exchanges on $\underset{\sim}{u}$ to obtain $\underset{\sim}{v}$ of the form $(r^{xa}, r^{-b}, s^{xc}, s^{-d})$ where $a, b, c, d$ are elements of the group $Q = gp\{r^x, s^x, r, s\} = gp\{xyx, s^{-1}y^{-1}x^2, x^2y, y^{-1}x\}$ $= gp\{x^3, x^{-1}y^{-1}x^2, x^2y\}$ . Now the centraliser $C(r) = gp\{r\} \subseteq Q$ and $Q$ does not contain $x$ . Hence $xQ \cap C(r) = \emptyset$ . But $r^{xa}$ cancels with $r^{-b}$ if and only if $xab^{-1} \in C(r)$ , which we have just shown impossible.

## 2. On Peiffer transformations, and link diagrams.

(Extract from a letter to Roger Fenn.)

The example given before, of a trivial identity sequence which does not collapse, arose in the following way.

The first example to try is $(x; x^2)$ , but this does not work on the algebra level. The point is you are working in $F = F(x, y, \ldots)$ and so $r^x$ and $r$ cannot be different if they are given by the same word. In fact any identity sequence $(r^u, s^v)$ $(r, s \in R,\ u, v \in F)$ collapses by definition of deletion: you are allowed to delete an adjacent pair which cancels in $F$ . So you need at least $(r^a, s^b, r^c, s^d)$ and you also need $r$ not conjugate to $s^{\pm 1}$ because Lyndon proved that a presentation with a *single* relator is necessarily aspherical in the strong sense that every identity among the conjugates of one relator actually collapses to $\emptyset$ . (This is a difficult theorem - see Proposition 10.6 and its proof on p. 160 and also the remarks preceding Proposition 11.1 on p. 161 in [L-S] [see also [C-C-H]].) Trying $(x; x^2) \cong (x, y; xy^{-1}, yx)$ fails on this conjugacy condition, so the next simplest thing is to try $(x; x^3) \cong$ $(x, y; x^2y, y^{-1}x)$ and that works.

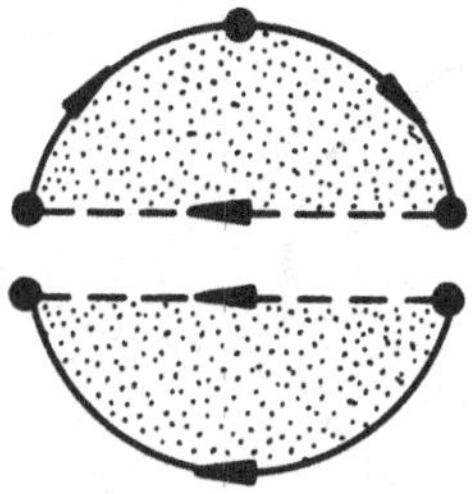

If I understand your letter correctly, $r^x$ and $r$ *can* represent a different element of $\pi_2 K$ even when equal in $F$, which shows that Lyndon's 'Peiffer machine' is a bad model for $\pi_2 K$. [This point is about the distinction between Y-sequences and $R^F$-sequences: see §6 of [Br-Hu]].

A bad thing about our examples is that they seem to depend heavily on the presence of torsion in

$$\pi_1 K = G \cong (X; R) .$$

On the other hand $\pi_2 K = 0 \Longrightarrow K = K(G, 1) \Longrightarrow$ cohomological dimension of a subgroup is less than or equal to that of the group, and $\mathbb{Z}_n$ has cohomological dimension $\infty$.

I like very much your description of homotopies in terms of bridge moves, but your strip in the first letter can be shortened a bit - only one insertion is needed:

These bridge moves occur in my Peiffer moves as follows:

$$(t^{-1}, r, s, t, s^{-1}, r^{-1}) \xmapsto{\text{(a and b)}} (t^{-1}, r, s, s^{-1}, r^{-1}, t^{\widetilde{\widetilde{sr}}}) \xmapsto{\text{collapse}} \emptyset .$$

I have a different way of drawing pictures of such homotopies which is simpler but much less useful as it does not carry as much information as your method. Essentially, one ignores the edges of your graphs and worries only about the vertices [i.e. the discs]. This is inspired by Whitehead's approach to $\pi_2(X, X_0)$ where $X = X_0 \cup$ (2-cells). A neat exposition of Whitehead's proof that $\pi_2(X, X_0)$ is a free crossed $\pi_1 X_0$-module has been written out by Ronald Brown [Br1]. Essentially, the construction is exactly the one explained to me by you and Colin Rourke, except that one does not worry about $K^1$ and looks only at the pre-image of the mid-points of the 2-cells. So the homotopy is represented by a kind

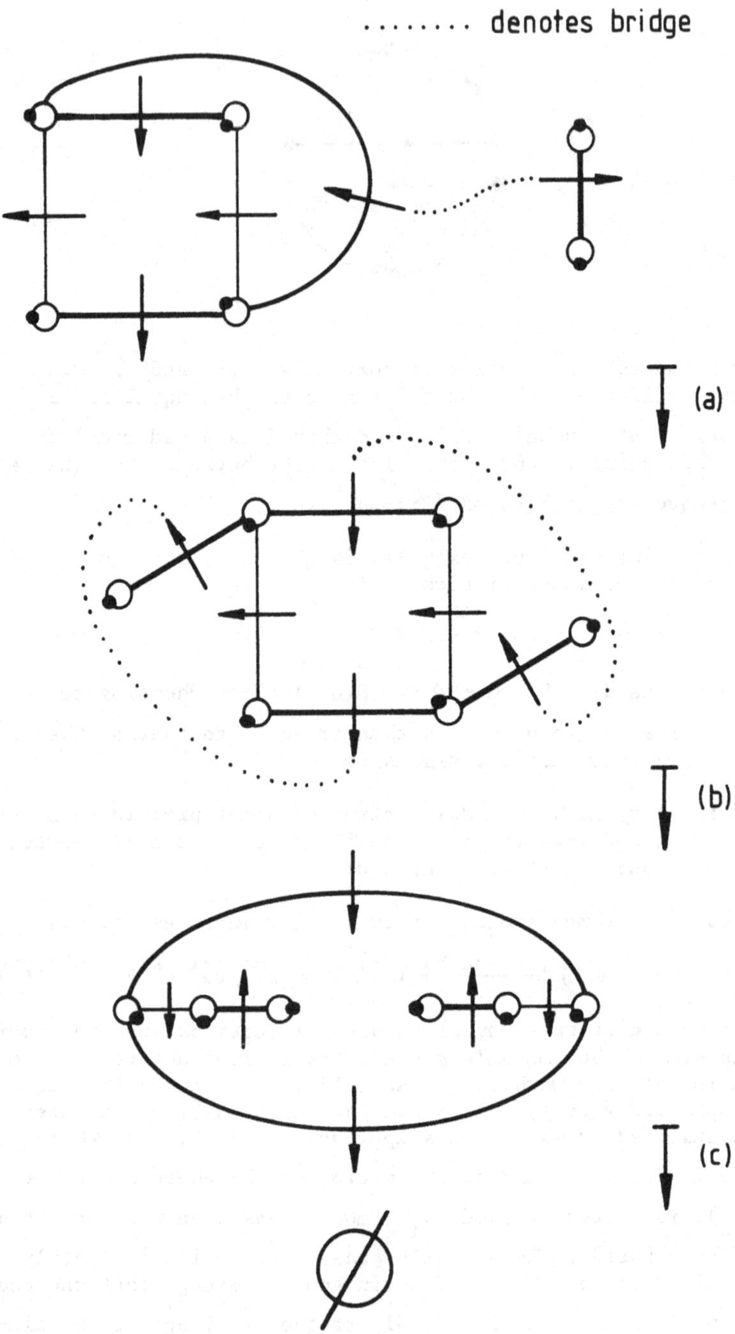
........ denotes bridge
(a)
(b)
(c)

of "linkage"; the relationship to $\pi_1$ is not visible, but somehow it did not matter to Whitehead.

There is a completely *formal* way of getting these linkages out of the Peiffer moves: the two exchanges are represented by

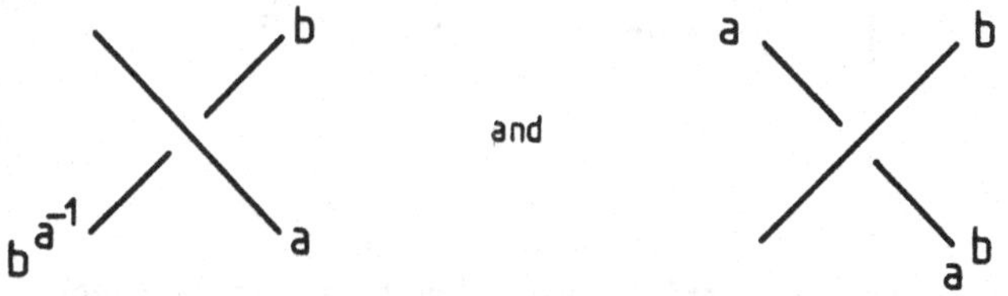


(the line in front represents the element which does not change). The insertions and deletions are represented by

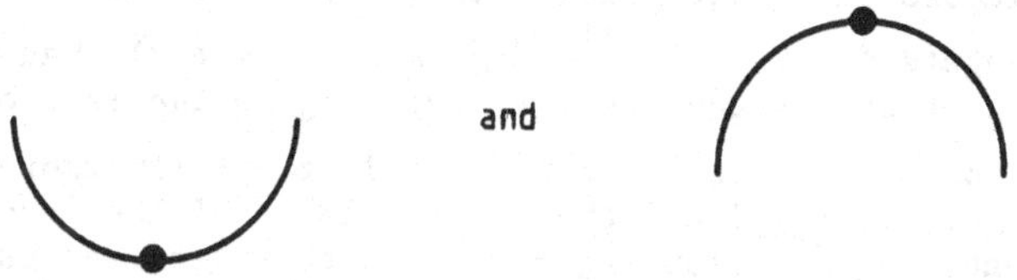


and one reads the diagram upwards. For example, the sequence $(r^x, s^x, r^{-1}) \longmapsto (t^{-1}, t, r^x, s^x, s^{-1}, r^{-1}) \longmapsto \ldots \longmapsto \emptyset$ of §1 is given by

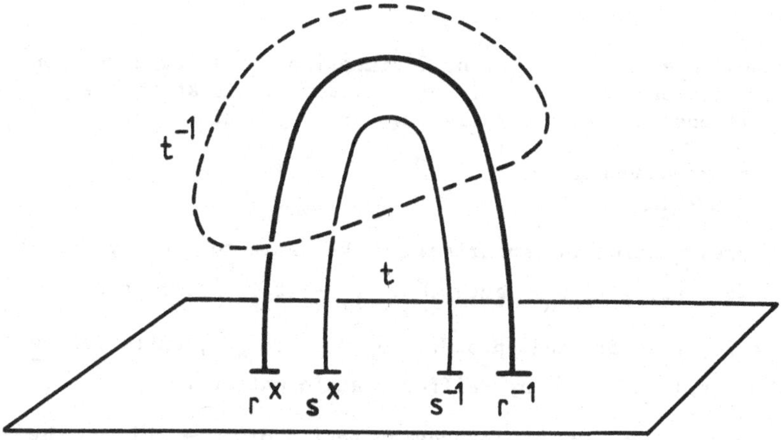

The obvious elementary moves are clearly allowed such as

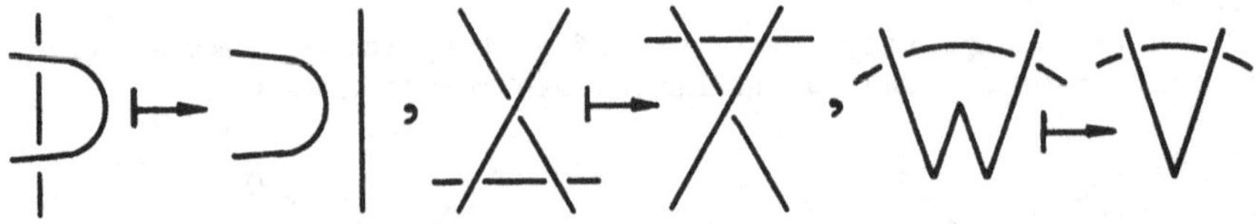

(This is obvious for geometric reasons, but it can be checked directly from the Peiffer transformations.) In contrast to your diagrams, the linkages capture only the formal side of things and are independent of the details of the presentation. In fact, some linkages are impossible - see below. If $R$ is *irredundant* (no $r \in R$ is conjugate in $F$ to $s^{\pm 1}$ for $s \in F$, $s \neq r$) then we may *colour* each strand in the linkage according to the relator $r$ whose conjugates $r^a = a^{-1}ra$, $a \in F$, label it. Furthermore, Peiffer exchanges clearly leave the class mod $N$ of the 'exponent' $a \in F$ invariant, so the colouring may be further refined: each basic colour (one for each $r \in R$) is subdivided into *shades* (one for each class mod $N$, i.e. one for each element of $G = F/N$). If, further, $F$ is *primary* (no $r \in R$ is a proper power of some element of $F$) then the centralizer of each $r \in R$ is simply the cyclic group generated by $r$ and so $C(r) \subset N$. Hence, *if* $R$ *is irredundant and primary, then only strands of the same shade can cancel.*

If now $K = |(X; R)|$ is the 2-dimensional complex corresponding to the presentation $(X; R)$ of $G$, then it is known that $\pi_2 K = 0$ if and only if the following three conditions hold:

1. $R$ is irredundant,
2. $R$ is primary,
3. The presentation is aspherical in the sense that every n-tuple $\underset{\sim}{p} = (p_1, p_2, \ldots, p_n)$ such that $p_i = (r_i^{\varepsilon_i})^{u_i}$, $r_i \in R$, $u_i \in F$, $\varepsilon_i = \pm 1$ and $p_1 p_2 \ldots p_n = 1$ in $F$, satisfies $\underset{\sim}{p}$ is equivalent to $\emptyset$ by Peiffer transformations.

Now conjugation in $F$ induces an action of $G = F/N$ on the abelianised group $\overline{N}$ making $\overline{N}$ into a $\mathbb{Z}G$-module. Let $\Delta_r : \mathbb{Z}G \longrightarrow \overline{N}$ be given by $x \longmapsto \overline{r}.x$ where $\overline{r}$ is the class of $r \in R$ in $\overline{N}$. The image of $\Delta_r$ is the cyclic submodule $\overline{r}.\mathbb{Z}G$

of $\overline{N}$ . Consider the homomorphism of $\mathbb{Z}G$-modules

$$\Delta = \bigoplus_{r\in R} \Delta_r : \bigoplus_{r\in R} \mathbb{Z}G \longrightarrow \overline{N} .$$

Then conditions 1, 2, and 3 above (and so the condition $\pi_2 K = 0$) are equivalent to the requirement that $\Delta$ is an isomorphism of $\mathbb{Z}G$-modules. It follows that if $\pi_2 K = 0$ then we have for every $r \in R$ a well defined homomorphism $Deg_r$ of $N$ onto the free abelian group $\mathbb{Z}G$ given by

$$Deg_r : N \xrightarrow{\lambda} \overline{N} \xrightarrow{\Delta^{-1}} \bigoplus_{r\in R} \mathbb{Z}G \xrightarrow{p_r} \mathbb{Z}G$$

where $\lambda$ is abelianisation and $p_r$ is the r'th projection. Note that $Deg_r$ is defined by the conditions

$$Deg_r((s^k)^a) = k\,[a]\,\delta(r, s)$$

for every $s \in R$ , $a \in F$ and integer $k$ ; $[a]$ denotes the image of $a \in F$ in $G = F/N$ and $\delta(r, s)$ is the Kronecker $\delta$ . Following $Deg_r$ by the evaluation homomorphism $\mathbb{Z}G \longrightarrow \mathbb{Z}$ at various points of $G$ or by the augmentation homomorphism $\mathbb{Z}G \longrightarrow \mathbb{Z}$ we get various integer valued 'degrees'.

Using these degrees, together with the fact that a centralizer of an element of a free group is necessarily a cyclic subgroup of the free group, we can show that certain linkages are forbidden in an aspherical $K(\pi_2 K = 0)$ .

EXAMPLE 1. *(Whitehead linkage)*. Suppose $K = |(X; R)|$ where $\pi_2 K = 0$ and $r, s \in R \subset F$ with $r \neq s$ . Let $a = (r^{\varepsilon})^u$ , $b = (s^{\eta})^v$ where $u , v \in F$ and $\varepsilon , \eta = \pm 1$ . Write $Deg_a$ for $\varepsilon.Deg_r : N \longrightarrow \mathbb{Z}G$ followed by the evaluation map $\mathbb{Z}G \longrightarrow \mathbb{Z}$ at $[u] \in G$ . Let $Deg_b$ be defined similarly in terms of $\varepsilon.Deg_s$ and $[v] \in G$ . Let $\tilde{c}$ denote $c^{-1}$ .

Suppose given a 'Whitehead linkage' corresponding to a sequence of Peiffer transformations. Starting at the top we find

$$c = a^{\tilde{b}} , d = b^c , e = c^{\tilde{a}} , f = d^{\tilde{e}} = (b^c)^{\tilde{c}^{\tilde{a}}} .$$

Now $f = b$ implies $c(\tilde{c}^{\tilde{a}})$ commutes with $b$ . As $b$ is a conjugate of $s \in R$ , and $R$ is primary, the centralizer of $b$ in $F$ is the cyclic group $< b >$ , and so

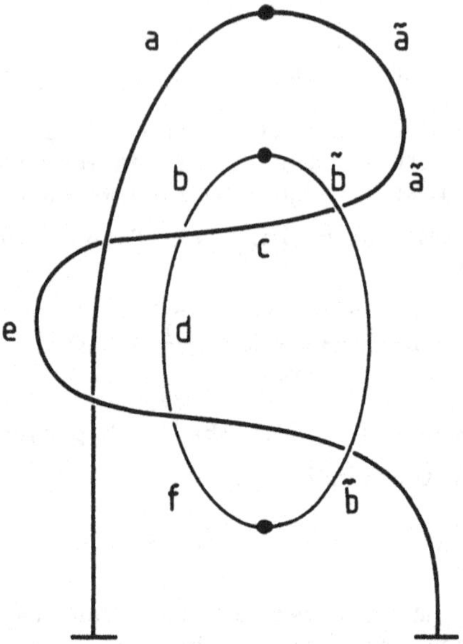


$$c(c^a) = b^k \qquad \text{for some } k \in \mathbb{Z},$$

$$\text{or} \quad (\tilde{a}^{\tilde{b}})(a^{\tilde{b}\tilde{a}}) = b^k .$$

Applying $\mathrm{Deg}_b$ to both sides of this equation, we obtain $k = 0$. Hence

$$(\tilde{a}^{\tilde{b}})(a^{\tilde{b}\tilde{a}}) = 1$$

$$\text{i.e.} \quad a^{\tilde{b}\tilde{a}} = a^{\tilde{b}}$$

$$\text{i.e.} \quad a^{\tilde{b}\tilde{a}b} = a .$$

So $\tilde{b}\tilde{a}b = \tilde{a}^b$ commutes with $a$, and so we obtain $\tilde{a}^b = a^\ell$ for some $\ell \in \mathbb{Z}$. Taking $\mathrm{Deg}_a$ of this equation, we obtain $\ell = -1$ and so $b$ commutes with $a$. Hence $b = a^m$ for some $m \in \mathbb{Z}$. Taking $\mathrm{Deg}_b$ we obtain $m = 0$, i.e. $b = 1$, which is a contradiction. Hence the *Whitehead linkage (with different coloured strands) never occurs if* $\pi_2 K = 0$.

EXAMPLE 2. Again assume that $\pi_2 K = 0$. Assume also that $x = (t^{\pm})^u$ where $t \in R$ and $r$, $s$ are conjugates of some relators *other than* $t$ (or conjugates of inverses of such relators). Again $rs$ must commute with $x$ and so $rs = x^k$.

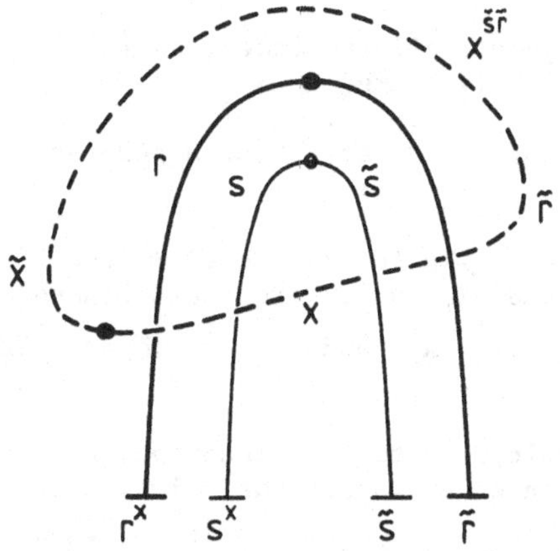


Taking $\mathrm{Deg}_x$ , we get $k = 0$ or $rs = 1$ . Hence the above diagram can be replaced by a simpler one

To prove (or to *disprove*) the Whitehead conjecture it is sufficient to consider the case when the subcomplex L of K differs from K by a single 2-cell. Paint this extra 2-cell red and all the remaining 2-cells blue. Assume that $\pi_2 K = 0$ and let $p = (p_1, p_2, \ldots, p_n)$ be an identity amongst the blue relators (each $p_i$ is a conjugate of a relator in L , or its inverse, and $p_1 p_2 \cdots p_n = 1$ in F) . Then p is equivalent to $\emptyset$ by Peiffer moves in K . This gives a linkage consisting of a blue part X anchored to the 'floor' and a red part, a link Y , floating above. To prove that $\pi_2 L = 0$ we must show that p is equivalent to $\emptyset$ in L - that is that we can get rid of the red stuff. (There is no need to show p collapses to $\emptyset$ - we are not worried about the blue insertions.)

If X and Y are geometrically unlinked, we are done. Otherwise, there seem to be two possible cases:

1. X and Y are algebraically linked as in Example 2. *Part of the problem is to make this idea precise* - perhaps in terms of the various degrees (shades of blue, shades of red).

2. X and Y are algebraically unlinked, as in Example 1, but are still geometrically linked.

Now a possible proof (or a search for a counter example) could go as follows (assuming $\pi_2 K = 0$):

(a) In case 1, one would like to show that X and Y can be replaced by $X_1$ and $Y_1$ which are 'less linked' algebraically, and so on, ending with $X_n$ and $Y_n$ which are algebraically unlinked.

(b) In case 2, one would like to show that this never happens (for $\pi_2 K = 0$): in other words, the only way to be algebraically unlinked in *aspherical* K is to be also geometrically unlinked.

Unfortunately, the 'degrees' described above do not seem a sufficient tool for all this - they work satisfactorily if there are only two underpasses, but after that they seem to fail. For example, I am unable to show that a 'double Whitehead link' cannot occur if $\pi_2 K = 0$:

It would help a lot if one had some 'higher order degrees' to decide whether a given identity sequence $p = (p_1, \ldots, p_n)$ is equivalent to $\emptyset$. Assume now that R is irredundant and primary, but do *not* assume that (X; R) is aspherical. The first (or zero-order) obstruction is of course $p_1 p_2 \cdots p_n = 1$ in F. Next, the first order obstruction is $\mathrm{Deg}_r p = 0$, for all $r \in R$. ($\mathrm{Deg}_r$ is not well-defined on N if (X; R) is not aspherical, but it is still well-defined on n-tuples and also on H (the free group on $F \times R$ , see [Br-Hu]).) If all $\mathrm{Deg}_r p = 0$ , it is still possible that p is not equivalent to $\emptyset$ , and it would be nice to have a second-order obstruction, defined when the first-order one is zero, and so on. It is all a bit reminiscent of higher order linking numbers, and Massey products have been suggested as being relevant.

Here are a few final points;

1. If $\pi_2 K = 0$ and if N is freely generated by a set of conjugates of elements of R , then every identity among relators actually collapses [L-S, p.160] and so every subcomplex of K is aspherical. Special cases of this are R 'staggered' and a special case of this is any one-relator presentation (p. 161). [See also [C-C-H].]

2. Other situations when $\pi_2 K = 0$ implies $\pi_2 L = 0$ for $L \subset K$ are listed by J.M. Cohen in [Co]. [See also [Hu1].] He also states two conjectures which would (together) imply the Whitehead conjecture. [But these conjectures have been shown false in [Be1] and [Du5].]

1 and 2 seem to indicate that a counter example to the Whitehead conjecture would have to be quite complicated, but that *could* be a false impression.

# Higher-dimensional group theory

R. BROWN

## 1. Introduction

The work I will be describing does hold in all dimensions. The reason for presenting it at a Conference on Low-Dimensional Topology is that the jump from one to two dimensions is a significant one, and that even in dimension two the applications are new.

The aim of the exposition is to give an idea of the main thrust of a series of papers [6 - 13] with Chris Spencer and with Philip Higgins which have developed and applied a new algebra of double groupoids and $\omega$-groupoids.

One justification of these new gadgets is that they occur, and in a non-trivial fashion, in both algebraic and geometric situations. They were sought out in order to formulate and prove a generalisation to all dimensions of the Seifert-van Kampen theorem on the fundamental group of a union of spaces, a generalisation which turned out to include as very special cases also the Brouwer degree theorem $(\pi_n(S^n) = \mathbb{Z})$, the relative Hurewicz theorem (in the form of a description of $\pi_n(X \cup CA)$), and the fact that $\pi_n(X \cup \{e^n_\lambda\}, X)$ is a free $\pi_1 X$-module for $n > 2$. These results are usually proved by classical tools of algebraic topology such as simplicial approximation, homology and covering spaces.

Another corollary of our main result is J.H.C. Whitehead's subtle theorem that $\pi_2(X \cup \{e^2_\lambda\}, X)$ is a free crossed $\pi_1 X$-module; Whitehead's proof uses methods of transversality and knot theory [25 - 27 and 5]. Listed in [6] are seven papers applying this theorem, which was also evidence for and a test case of the two-dimensional Seifert-van Kampen theorem of [6]. At the present time, no text book on algebraic topology contains a proof of Whitehead's theorem, whose difficulty seems to be that it gives information (as does the Seifert-van Kampen theorem) on a non-abelian group.

At first sight it would seem that the Seifert-van Kampen theorem has little in common with the other results mentioned. However all these results assert that a given group (or module) has a universal property; it is desirable that they should all be proved by verifying the universal property. This aim is achieved, and it perhaps explains why our proofs use none of the tools mentioned above.

Instead we attempt to rewrite aspects of homotopy theory by developing algebraic structures which model the geometry more closely than the usual group or module structures. In particular, we require structures of a group-like character but suitable for dealing with techniques of subdivisions and compositions of cubes. Since these techniques are basic in many areas of mathematics, it is possible that this algebra will have wider application than the elementary aspects of homotopy theory discussed here.

## 2. The fundamental groupoid

We start with the generalisation from the fundamental group $\pi_1(X, x_0)$ of a space with base point to the fundamental groupoid $\pi_1(X, X_0)$ of a space with subset $X_0$. Here $X_0$ is thought of as a set of base points and $\pi_1(X, X_0)$ is defined as the set of homotopy classes rel $\dot{I}$ of maps $(I, \dot{I}) \to (X, X_0)$ with groupoid structure induced by the usual composition of paths (written as $a + b$).

It was known for a long time that the fundamental groupoid allowed for a conceptualisation of change of base point. The further observation made in [3] (stimulated by work of Philip Higgins) was that the fundamental groupoid allows for a generalisation of the Seifert-van Kampen theorem to non-connected space, and that this theorem is useful for computations.

THEOREM 1 [3]. *Let* $U, V$ *be subsets of* $X$ *whose interiors cover* $X$. *Let* $X_0$ *be a subset of* $X$ *meeting each path-component of* $U, V$ *and* $W = U \cap V$. *Let* $U_0 = U \cap X_0$, $V_0 = V \cap X_0$, $W_0 = W \cap X_0$. *Then the diagram of morphisms induced by inclusion*

$$\begin{array}{ccc} \pi_1(W, W_0) & \xrightarrow{a} & \pi_1(U, U_0) \\ {\scriptstyle b}\downarrow & & \downarrow{\scriptstyle \bar{b}} \\ \pi_1(V, V_0) & \xrightarrow[\bar{a}]{} & \pi_1(X, X_0) \end{array}$$

*is a pushout in the category of groupoids.*

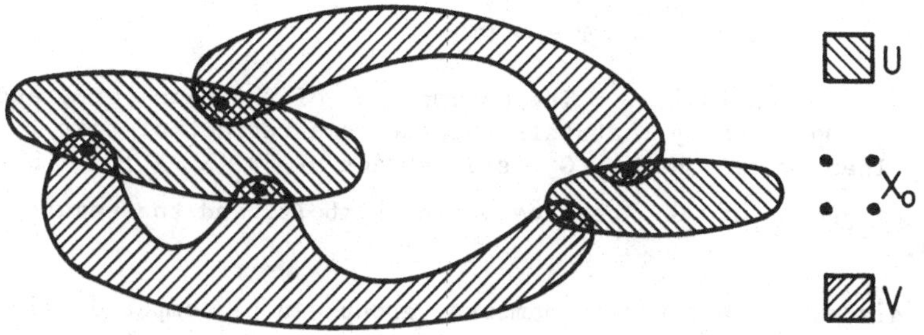


Fig. 1

The above picture shows the type of application that can be handled. The determination of each fundamental group

$$\pi_1(X, x_0) \ , \ x_0 \in X_0 \ ,$$

using the above theorem is a purely algebraic matter involving presentations of groupoids. There are several ideas in this which do not occur in group theory, for example the universal groupoid $U_\sigma(G)$ "induced" from a groupoid $G$ by a function $\sigma : \mathrm{Ob}G \to S$ (a full account is given in [4 , 20]). Intuitively, presentations of groupoids require discussion of relations in dimension 0 and dimension 1 , and this corresponds nicely to the geometry.

The following Corollary of Theorem 1 is one example of results which are more troublesome to prove using groups alone. (See also [4] p.289, Ex.4).

COROLLARY *Let* $U, V$ *be path-connected open subsets of* $X$ *such that* $X = U \cup V$ *, and suppose that* $W = U \cap V$ *has* $n + 1$ *path-components. Let* $x_0 \in W$ *. Then* $\pi_1(X, x_0)$ *contains the free group on* $n$ *symbols as a retract.*

*Proof.* Let $X_0$ be such that $x_0 \in X$ and $X_0$ meets each path-component of $W$ in a single point. Let $D$ be the discrete groupoid on $X_0$ (i.e. $D$ consists only of identities). Let $T$ be the tree groupoid on $X_0$ (i.e. $T(x, y)$ has one element for all $x, y \in X_0$). Form the pushout of groupoids

$$\begin{array}{ccc} D & \xrightarrow{i} & T \\ {\scriptstyle i}\downarrow & & \downarrow \\ T & \longrightarrow & G \end{array}$$

in which $i$ is inclusion. The assumptions of the Corollary, and Theorem 1, imply easily that this pushout is a retract of that given by Theorem 1. Hence $G$ is a retract of $\pi_1(X, X_0)$. But the vertex groups of $G$ are free on $n$ symbols, and the result follows. □

The discovery that the fundamental group can be computed using groupoid methods suggested the possibility of exploiting these methods in other areas of homotopy theory. Applications to covering spaces and a gluing theorem for homotopy equivalences were given in [4]; for other applications, see [19].

The generalisations of the Seifert-van Kampen theorem referred to in §1 are perhaps the first applications of groupoids really to touch problems of higher homotopy groups. However, the techniques are new and elaborate. In order to explain and motivate the key ideas we give a sketch proof of Theorem 1 in a form and notation suitable for generalisation - the proof is essentially that of Crowell in [14] for the fundamental group.

## 3. Proof of Theorem 1

Suppose given a commutative square of morphisms of groupoids

$$\begin{array}{ccc} \pi_1(W, W_0) & \longrightarrow & \pi_1(U, U_0) \\ {\scriptstyle b}\downarrow & & \downarrow{\scriptstyle b'} \\ \pi_1(V, V_0) & \xrightarrow[a']{} & G \end{array}$$

We have to prove there is a unique morphism $f : \pi_1(X, X_0) \to G$ such that $f\bar{a} = a'$, $f\bar{b} = b'$. This is done by constructing elements $F\alpha$ of $G$ for $\alpha : (I, \dot{I}) \to (X, X_0)$.

Suppose first given a path $\theta$ in $X$ such that $\theta$ has a subdivision $\theta = \theta_1 + \ldots + \theta_n$ such that each $\theta_i$ is a map $(I, \dot{I}) \to (X, X_0)$ and lies in $U$ or $V$. Then it is easy to use the morphisms $a'$, $b'$ and the condition $a'b = b'a$ to determine elements $F\theta_i$ which compose in $G$ to give an element $F\theta_1 + \ldots + F\theta_n$ which we write as $F\theta$ although *a priori* it depends on the subdivision chosen.

We next wish to construct $F\alpha$ in $G$ for an arbitrary path $\alpha : (I, \dot{I}) \to (X, X_0)$, and to prove $F\alpha$ is independent of the choices involved. For this purpose we use the following lemma for the cases $m = 1, 2$.

LEMMA 1. *Let* $\alpha : I^m \to X$ *be a map of a cube such that* $\alpha$ *maps the set* $I_0^n$ *of vertices of* $I^m$ *into* $X_0$. *Suppose that* $I^m$ *is subdivided (by planes parallel to* $x_i = 0$, $i = 1,\ldots,m$ *) into subcubes* $c_\lambda$ *and that for all* $\lambda, \alpha(c_\lambda)$ *lies in* $U$ *or in* $V$. *Then there is a homotopy* $h : \alpha \simeq \theta$ rel $I_0^m$ *such that*

*(i)* *if* $\alpha(c_\lambda)$ *lies in* $U$ *(or* $V$ *) so also does* $h(c_\lambda \times I)$ *;*

*(ii)* $\theta$ *maps the vertices of* $c_\lambda$ *into* $X_0$ *;*

*(iii)* *if* $e^r$ *is a face of* $c_\lambda$ *and* $\alpha(e^r) \subset X_0$, *then* $h(e^r \times I) \subset X_0$. □

The proof is an easy induction on the skeleta of the cell structure $K$ determined by the subdivision of $I^m$.

COROLLARY. *If* $\alpha : (I, \dot{I}) \to (X, X_0)$ *is any path, then there is a homotopy* $h : \alpha \simeq \theta$ rel $\dot{I}$ *such that* $F\theta$ *is defined.* □

The proof is obtained by constructing a subdivision $\alpha = \alpha_1 + \ldots + \alpha_n$ such that each $\alpha_i$ lies in $U$ or in $V$, and applying Lemma 1 with $m = 1$.

Given $\alpha$ as in the Corollary, we choose $\theta$ so that $F\theta$ is defined and set $F\alpha = F\theta$. We now want to prove $F\alpha$ depends only on the equivalence class of $\alpha$ in $\pi_1(X, X_0)$. For this it is sufficient to prove $F\theta = F\theta'$ whenever there is a homotopy $h : \theta \simeq \theta'$ rel $\dot{I}$ and the elements $F\theta$, $F\theta'$ of $G$ are determined by subdivisions $\theta = \theta_1 + \ldots + \theta_n$, $\theta' = \theta'_1 + \ldots + \theta'_n$ such that each $\theta_i$, $\theta'_j$ lies in $U$ or in $V$ and maps $\dot{I}$ into $X_0$.

Now $h$ is a map $I^2 \to X$. We subdivide $I^2$ as in Lemma 1 into subsquares $c_\lambda$ such that $h(c_\lambda)$ lies in $U$ or in $V$. This subdivision may be further refined so that it refines on $I \times \{0\}$ and $I \times \{1\}$ the given subdivisions of $\theta, \theta'$. Then Lemma 1 can be applied to deform $h$ to a homotopy

$$k : \phi \simeq \phi' \text{ rel } \dot{I}$$

where $F\phi, F\phi'$ are defined and satisfy $F\theta = F\phi$, $F\theta' = F\phi'$.

Thus we have to prove $F\phi = F\phi'$. This is the key part of the proof, and it was an analysis of this part which suggested that higher dimensional analogues should be possible.

LEMMA 2. *Let* $k : \phi \simeq \phi'$ *be a homotopy* rel $\dot{I}$ *of paths* $\phi, \phi' ; (I, \dot{I}) \to (X, X_0)$ *and suppose* $I^2$ *has a subdivision into squares* $c_{ij}$ *such that*

*(i)* $k$ *maps all the vertices of the subdivision into* $X_0$,

*(ii)* $k(c_{ij})$ *lies in* $U$ *or* $V$.

*Let* $F\phi, F\phi'$ *be defined by the induced subdivisions on* $I \times \{0\}$, $I \times \{1\}$. *Then* $F\phi = F\phi'$. □

The proof of Lemma 2 is of a standard type (for details see [3, 14]) and here we give only the key part of the argument in simplified form as follows.

Suppose given two maps of squares into $X$ which lie in $U$ or in $V$ and whose edges define paths in $X$ with end points in $X_0$, as shown below.

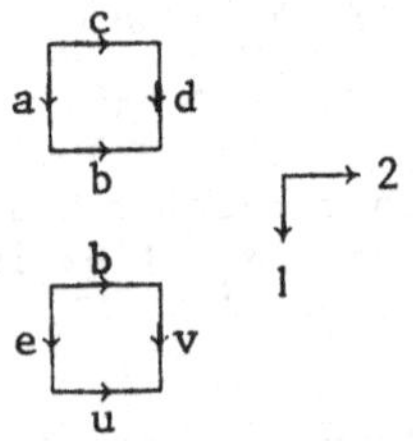


Then $d \simeq -c + a + b$, $v \simeq -b + e + u$ in $U$ or in $V$ so that in the groupoid $G$ we have

$$Fd = -Fc + Fa + Fb \;,\; Fv = -Fb + Fe + Fu$$

whence

$$\begin{aligned} Fd + Fv &= -Fc + Fa + Fb - Fb + Fe + Fu \\ &= -Fc + Fa + Fe + Fu. \end{aligned}$$

Thus the contribution from b has cancelled.

Once Lemma 2 has been proved, the remaining verifications in the proof of Theorem 1 are simple.

REMARK. A somewhat different proof of Theorem 1 is given in [3, 4] in that the result is first proved for the case $X_0 = X$, and the general case is obtained by a retraction argument. Unfortunately, this appealing argument does not seem to generalise either to the case of arbitrary unions, or to higher dimensions.

Best possible results on the fundamental groupoid are given in [23].

## 4. Double groupoids

We now begin to show how to mimic the above proof in one higher dimension, using maps of $I^2$ instead of $I^1$. At this stage there is an idea of a proof, in search of a theorem.

There are two clues as to how to proceed. First, in view of the success of groupoids in dimension one, we look for groupoid rather than group structures. Second, in groupoids, as in groups, compositions are on one line as in $g_1 \dots g_n$, whereas we need *arrays* in order to model subdivision of a square:

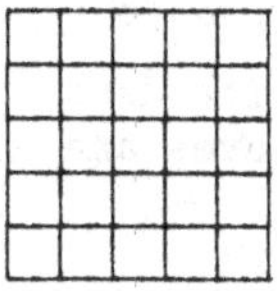

Fig. 2

So we look for some kind of double groupoid.

Double groupoids have been considered by Ehresmann and others as sets with two mutually compatible groupoid structures (alternatively, as groupoid objects in the category of groupoids). The geometry suggests a more restrictive definition, which is best introduced after a precise definition of groupoid.

A *groupoid* G consists of a set $G_0$ of points (or *vertices*) and a set $G_1$ of *edges* together with the following structures.

(i) Maps $\partial^0, \partial^1 : G_1 \to G_0$ and $\varepsilon : G_0 \to G_1$ called the *initial, final,* and *identity* (or *degeneracy*) maps.

(ii) A partial composition of edges $(a, b) \mapsto a + b$ defined if and only if $\partial^1 a = \partial^0 b$.

(iii) An involution $a \mapsto -a$ on $G_1$.

These shall satisfy the rules : $\partial^0 \varepsilon = \partial^1 \varepsilon = \mathrm{id}$; + is associative; $\varepsilon\partial^0 a + a = a = a + \varepsilon\partial^1 a$ ; $\partial^0(-a) = \partial^1 a$, $\partial^1(-a) = \partial^0 a$ ; $-a + a = \varepsilon\partial^1 a$, $a + (-a) = \varepsilon\partial^0 a$.

A *double groupoid* G consists of sets $G_0$ of points, $G_1$ of *edges*, $G_2$ of *squares* with boundary maps

$$G_2 \begin{smallmatrix}\longrightarrow\\ \longrightarrow\\ \longrightarrow\\ \longrightarrow\end{smallmatrix} G_1 \overset{\partial^1}{\underset{\partial^0}{\rightrightarrows}} G_0$$

and degeneracy maps

$$G_2 \leftleftarrows G_1 \xleftarrow{\varepsilon} G_0$$

satisfying the usual rules for the two dimensional part of a cubical complex. On $(G_1, G_0)$ is imposed a groupoid structure as above.

On $(G_2, G_1)$ we impose two groupoid structures $\underset{1}{+}, \underset{2}{+}$. It would be out of place here to write down all the rules involved - the main idea is that the structures are given by "vertical" and "horizontal" composition

$$\alpha \underset{1}{+} \beta \quad \begin{array}{|c|}\hline \alpha \\ \hline \beta \\ \hline \end{array} \qquad \begin{array}{|c|c|}\hline \alpha & \gamma \\ \hline \end{array} \; \alpha \underset{2}{+} \gamma$$

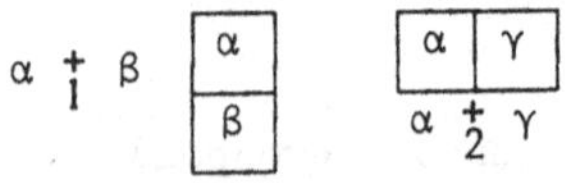

which are defined under the "obvious" boundary conditions and satisfy all the "obvious" laws, for example those describing the boundary edges of $\alpha \underset{1}{+} \beta$, $\alpha \underset{2}{+} \gamma$. The interaction of the two operations is given by the interchange law which says that

$$(\alpha \underset{1}{+} \beta) \underset{2}{+} (\gamma \underset{1}{+} \delta) = (\alpha \underset{2}{+} \gamma) \underset{1}{+} (\beta \underset{2}{+} \delta) \qquad (1)$$

whenever both sides are defined. This law is most simply written in matrix notation as giving a unique composite of

$$\begin{bmatrix} \alpha & \gamma \\ \beta & \delta \end{bmatrix}$$

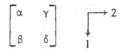

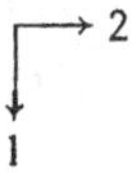


where in writing such a matrix we will always suppose the individual composites $\alpha \underset{1}{+} \beta$ , $\gamma \underset{1}{+} \delta$ , $\alpha \underset{2}{+} \gamma$ , $\beta \underset{2}{+} \delta$ are defined. This notation and convention is easily extended to $m \times n$ matrices defining arbitrary compositions.

At this stage a crucial difference emerges between group and groupoid theory. It is well known that if $\underset{1}{+}$ , $\underset{2}{+}$ are two *group* structures on a set satisfying (1), then they coincide and are abelian. The argument is simple and is worth giving here, since it is often used to explain why the second homotopy group $\pi_2(X,x)$ is abelian.

Suppose then $\underset{1}{+}$ , $\underset{2}{+}$ are group structures with zeros $0_1$ , $0_2$ . From (1)

$$0_1 \underset{2}{+} 0_1 = (0_1 \underset{2}{+} 0_1) + (0_1 \underset{2}{+} 0_1)$$

which implies $0_1 \underset{2}{+} 0_1 = 0_1$ , and so $0_1 = 0_2$ . We now write $0$ for $0_1$ . Next put $\beta = \gamma = 0$ in (1) to obtain

$$\alpha \underset{2}{+} \delta = \alpha \underset{1}{+} \delta$$

and $\alpha = \delta = 0$ in (1) to obtain

$$\alpha \underset{2}{+} \gamma = \gamma \underset{1}{+} \alpha .$$

Thus $\underset{1}{+}$ , $\underset{2}{+}$ coincide and are abelian. However no such deduction is possible in the groupoid case, since not all compositions are defined. One can prove only that $G_2$ contains a family of abelian groups.

REMARK. The last argument is often stated as: group objects in the category of groups are abelian groups. However, group objects in the category of groupoids are much more complicated - they are equivalent to crossed modules (for which see below). This result of Verdier is outlined and applied in [12].

The above algebra of double groupoids does not have quite enough structure. To see why this is so, we define the principal geometric example [6] and consider its possible use in generalising Theorem 1.

Let $\underset{\sim}{X} = (X, X_1, X_0)$ be a triple of spaces. Let $\rho\underset{\sim}{X}$ be the 2-dimensional cubical complex with $\rho_0\underset{\sim}{X} = X_0$, $\rho_1\underset{\sim}{X} = \pi_1(X_1, X_0)$ and $\rho_2\underset{\sim}{X}$ equal to the set of homotopy classes rel $\ddot{I}^2$ (where $\ddot{I}^2$ is the set of vertices of $I^2$) of maps $(I^2, \dot{I}^2, \ddot{I}^2) \to (X, X_1, X_0)$ the boundary maps and degeneracy maps are induced in the usual way. The surprising fact is that $\underset{1}{+}, \underset{2}{+}$, the usual compositions of squares in the two directions of $I^2$, induce compositions $\underset{1}{+}, \underset{2}{+}$ on $\rho_2\underset{\sim}{X}$ so that $\rho\underset{\sim}{X}$ *has the structure of double groupoid as defined above.* We call $\rho\underset{\sim}{X}$ the *homotopy double groupoid* of $\underset{\sim}{X}$.

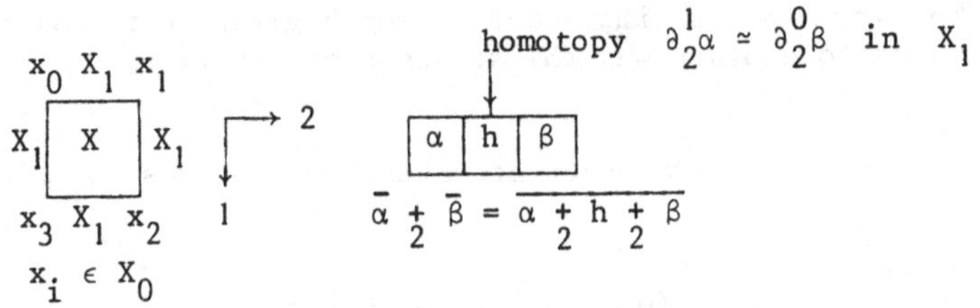


Fig. 3

The aim now is to formulate and prove a pushout theorem for $\rho\underset{\sim}{X}$, analogous to Theorem 1. So we suppose $X = U \cup V$, where $U, V$ are open, and set $W = U \cap V$. Let $\underset{\sim}{U}, \underset{\sim}{V}, \underset{\sim}{W}$ be the triples induced by intersection from $\underset{\sim}{X}$. Then there is a commutative diagram induced by inclusion

$$\begin{array}{ccc} \rho\underset{\sim}{W} & \longrightarrow & \rho\underset{\sim}{U} \\ \downarrow & & \downarrow \\ \rho\underset{\sim}{V} & \longrightarrow & \rho\underset{\sim}{X} \end{array}$$

and we ask: is this a pushout of double groupoids under reasonable conditions?

The answer turns out to be no, if double groupoids have only the structure discussed so far.

The difficulties in generalising the proof of Theorem 1 are easy to overcome up to the paragraphs before Lemma 2. All that is needed are the assumptions that $X_0$ meets each path-component of $U_0, U_1, V_0, V_1, W_0$ and $W_1$, and that $(U, U_1)$, $(V, V_1)$, $(W, W_1)$ are 1-connected. The real difficulties occur with the proof of Lemma 2, and the solutions of these difficulties gives a new twist to the algebra.

The form of the sketch argument for Lemma 2, but in one higher dimension is as follows. Suppose given two maps of cubes into $X$, whose vertices lie in $X_0$, whose edges lie in $X_1$, and which themselves lie in $U$ or in $V$. Suppose adjacent faces of these cubes coincide, as below.

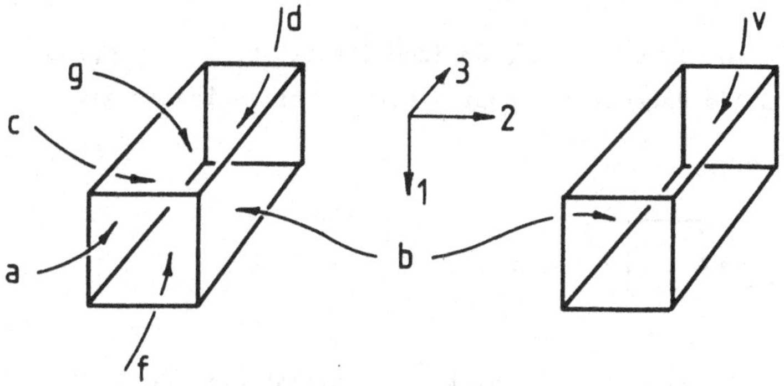


Fig. 4

Then the back faces $d$, $v$ should be expressable up to homotopy in terms of the other five faces of each cube, and so there should be expressions in $G$ (a double groupoid) for $Fd$, $Fv$. These should lead to an expression for $Fd +_2 Fv$ in which the contribution from $b$ has cancelled.

The expression of one face of a cube in terms of the other five is known as the *homotopy addition lemma*. In the next section we define the extra structure on double groupoids needed to formulate this lemma.

## 5. Thin elements

In order to define the extra structure needed, we consider the simplest example of double groupoid, due to Ehresmann [16].

Let $B$ be a groupoid. The double groupoid $D = \square B$ of *commuting squares* in $B$ has $D_0 = B_0$, $D_1 = B_1$ and $D_2$ consists of quadruples $\begin{pmatrix} a & {}^{c}_{b} & d \end{pmatrix}$ of elements of $B$ such that $-b-a+c+d$ is defined and is an identity. The compositions $+_1$, $+_2$ are then the obvious vertical and horizontal compositions of commuting squares.

Suppose given a double groupoid $G$ . Then there will be squares of $G$ with commuting boundary, that is with edges given by

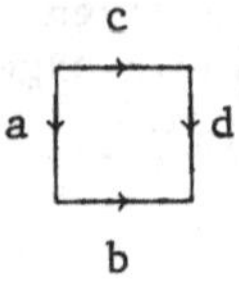


and for which $-b-a+c+d = 0$ , an identity of $G_1$ . Examples of such squares are degenerate squares with edges typically given by

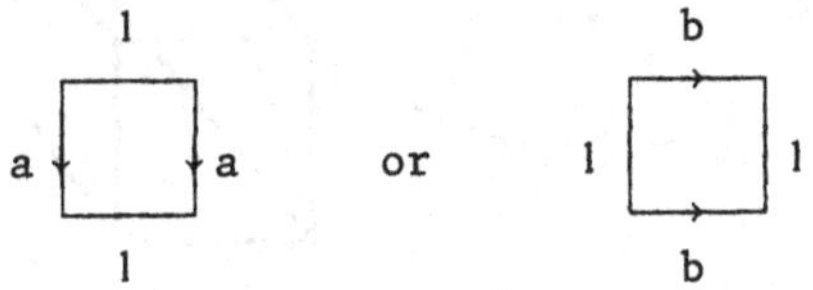


Among the others there seems no way to distinguish any one from another.

On the other hand, our geometric example $\rho\underset{\sim}{X}$ does have "special" squares with commuting boundary, namely those representable by maps $f$ with $f(I^2) \subset X_1$ . We call such elements of $\rho_2\underset{\sim}{X}$ *thin*. Thin squares of $\rho\underset{\sim}{X}$ are not necessarily degenerate - a typical example is obtained by retracting $I^2$ onto three sides and mapping these into $X_1$ with vertices in $X_0$ .

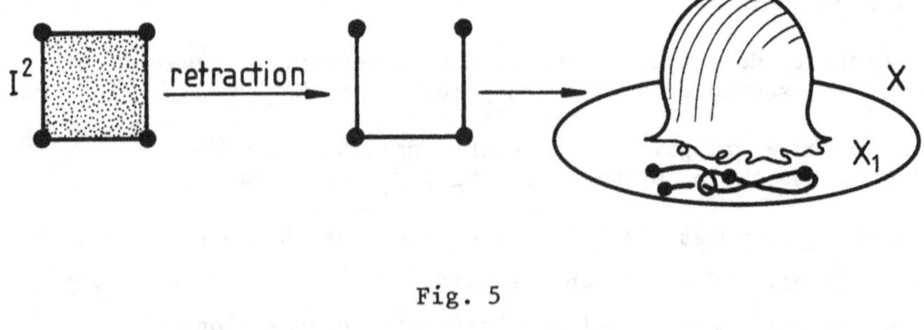


Fig. 5

We therefore impose on $G$ an additional structure of "thin" squares. This shall consist of a morphism

$$\Theta : \square G_1 \to G$$

of double groupoids which is the identity on $G_0$ and $G_1$. Of course a morphism of double groupoids simply preserves all the structure which is there, so that for example $\Theta\begin{pmatrix} & c & \\ a & & d \\ & b & \end{pmatrix}$ has the same edges as $\begin{pmatrix} & c & \\ a & & d \\ & b & \end{pmatrix}$. We call $\Theta\begin{pmatrix} & c & \\ a & & d \\ & b & \end{pmatrix}$ a *thin* square in $G$. That $\Theta$ is a morphism of double groupoids then implies: *any composite of thin squares is thin.*

From now on we assume that any double groupoid has this additional structure of thin squares.

Thin squares should be thought of as generalisations of identity elements in a groupoid. However there is a greater variety variety of thin squares than of identities and to handle them we introduce the following suggestive and convenient notation.

Thin squares with identities as 2 or 4 of their edges are written as one of

$$\| \quad = \quad \lrcorner \quad \urcorner \quad = \quad \ulcorner \quad \llcorner \quad \square \qquad (2)$$

the convention being that the lines denote identity edges. The other two edges in the first 6 cases are not given but they are related since a thin square has commuting boundary. Also a thin square is determined by its boundary edges. This shows that the first six squares in (2) are determined by one non-identity edge. It also allows us to write equations such as

$$\begin{bmatrix} \ulcorner & = & \lrcorner \\ \| & \square & \square \end{bmatrix} = \| \ , \quad \begin{bmatrix} \lrcorner & \| \\ = & \lrcorner \end{bmatrix} = \lrcorner$$

and

$$\begin{bmatrix} \ulcorner & = & \urcorner \\ \| & \square & \| \end{bmatrix} = \square \ ,$$

since we know that any composite of thin squares is thin, and the type of thin square from (2) given by a composite is determined by which edges are identities.

A slightly harder use of thin squares is given by "rotations"

$\sigma$ and $\tau$. To each square $\alpha$ in $G_2$ we assign squares

$$\sigma(\alpha) = \begin{bmatrix} = & \urcorner & \parallel \\ \ulcorner & \alpha & \lrcorner \\ \parallel & \llcorner & = \end{bmatrix}, \quad \tau(\alpha) = \begin{bmatrix} \parallel & \ulcorner & = \\ \llcorner & \alpha & \urcorner \\ = & \lrcorner & \parallel \end{bmatrix}.$$

We now use some repartitioning to compute

$$\tau\sigma(\alpha) = \left[\begin{array}{c|ccc|c} \parallel & \square & \square & \ulcorner & = \\ \hline \llcorner & = & \urcorner & \parallel & \square \\ \square & \ulcorner & \alpha & \lrcorner & \square \\ \square & \parallel & \llcorner & = & \urcorner \\ \hline = & \lrcorner & \square & \square & \parallel \end{array}\right] = \left[\begin{array}{ccc|cc} \parallel & \square & \square & \ulcorner & = \\ \llcorner & = & \urcorner & \parallel & \square \\ \hline \square & \ulcorner & \multicolumn{1}{|c|}{\alpha} & \lrcorner & \square \\ \hline \square & \parallel & \multicolumn{1}{|c}{\llcorner} & = & \urcorner \\ = & \lrcorner & \multicolumn{1}{|c}{\square} & \square & \parallel \end{array}\right]$$

$$= \left[\begin{array}{c|c|c} \square & \parallel & \square \\ \hline = & \alpha & = \\ \hline \square & \parallel & \square \end{array}\right] = \alpha .$$

Similarly, $\sigma\tau(\alpha) = \alpha$. Other formulae which may be proved are:

$$\sigma(\alpha \underset{1}{+} \beta) = \sigma(\alpha) \underset{2}{+} \sigma(\beta), \quad \sigma(\alpha \underset{2}{+} \gamma) = \sigma(\gamma) \underset{1}{+} \sigma(\alpha),$$

$$(\underset{2}{-} \sigma\alpha) \underset{1}{+} \tau\alpha = 0, \quad \sigma^2(\alpha) = \underset{2}{-}\underset{1}{-} \alpha, \quad \alpha^4(\alpha) = \alpha,$$

where $\underset{1}{-}, \underset{2}{-}$ denote the inverses for the groupoid structures $\underset{1}{+}, \underset{2}{+}$. These relations show that one effect of imposing the additional structure of thin squares is to ensure that the two groupoid structures $\underset{1}{+}, \underset{2}{+}$ are isomorphic.

All the above results apply to $\rho\underset{\sim}{X}$ and imply the existence of specific filter homotopies.

This structure of thin squares now allows for an expression for the homotopy addition lemma. The required formula is, with the notation of Fig. 4,

$$d \approx \begin{bmatrix} \ulcorner & {}^{-}_{1}\tilde{g} & \urcorner \\ {}^{-}_{2}c & a & b \\ \llcorner & \tilde{f} & \lrcorner \end{bmatrix}$$

in which $\tilde{f}$ is given by $(x, y) \mapsto f(y, x)$ . (The proof of this is elementary and easy, see [6] Proposition 5.) The required cancellation argument required for generalising the proof of Lemma 2 is then the following type of computation in the double groupoid G :

$$\begin{bmatrix} \ulcorner & {}^{-}_{1}F\tilde{g} & \urcorner & \ulcorner & {}^{-}_{1}F\tilde{v} & \urcorner \\ {}^{-}_{2}Fc & Fa & Fb & {}^{-}_{2}Fb & Fe & Fu \\ \llcorner & F\tilde{f} & \lrcorner & \llcorner & F\tilde{w} & \lrcorner \end{bmatrix} = \begin{bmatrix} \ulcorner & {}^{-}_{1}F\tilde{g} & {}^{-}_{1}F\tilde{v} & \urcorner \\ {}^{-}_{2}Fc & Fa & Fe & Fu \\ \llcorner & F\tilde{f} & F\tilde{w} & \lrcorner \end{bmatrix}$$

(In fact $F\tilde{f} = {}^{-}_{1}\,\sigma(Ff)$ .)

It is hoped that this will convince the reader of the validity of a two-dimensional version of Theorem 1. A precise statement and proof of a more general union theorem is given in [6]. An n-dimensional version of Theorem 1 is stated in §7.

REMARK. The above definition of double groupoid is equivalent to that of special double groupoid with special connection as given in [13] and called simply double groupoid in [6]. (The harder part of this equivalence is Proposition 1 of [6].) The exposition on rotation is due to Brown-Higgins. Related results on double categories are given in [24].

## 6. Interpretation

The previous two sections suggest two important questions. How do double groupoids arise algebraically? What use is a pushout theorem for $\rho\underset{\sim}{X}$ ? These questions are intimately related. Let us consider the first one.

We have already considered the double groupoid $D = \square B$ of commuting squares in a groupoid B . However, a homotopy theorist is unlikely to expect all squares to commute. So a natural generalisation is to suppose given a subgroupoid A of B and to change D to D' where $D'_1 = B$ but $D'_2$ consists of quadruples $\begin{pmatrix} a & {c \atop b} & d \end{pmatrix}$ of elements of B such that $\alpha = -b-a+c+d$

is defined and belongs to $A$. In order to have composition of squares well defined we must, as is easy to check, require $A$ to be normal in $B$, that is, if $\beta \in A(x,y)$, $\alpha \in A(x,x)$, then $-\beta+\alpha+\beta \in A(y,y)$. Also for this construction to work, we can assume without loss that $A$ is totally disconnected (i.e. $A(x,y) = \emptyset$ if $x \neq y$). The thin squares of $D'$ are the commuting squares.

The elements of $D'_2$ are determined by their boundary edges, and this too seems unreasonable. A natural generalisation is to suppose given a morphism $\delta : A \to B$ of groupoids where $A$ is totally disconnected, $Ob(A) = Ob(B)$ and $Ob(\delta)$ is the identity. We then change $D'$ to $D''$ where $D''_1 = B$ but $D''_2$ consists of quintuples $(\alpha\ ;\ a\,{}^{c}_{b}\,d)$ such that $a, b, c, d \in B$, $\alpha \in A$ and $-b-a+c+d = \delta(\alpha)$. In order to define $+_1$ and $+_2$ we also require $B$ to operate on $A$ on the right so that

$$\text{(i)} \qquad \delta(\alpha^a) = -a+\delta\alpha+a\ ,\ \alpha \in A\ ,\ a \in B\ .$$

The operations $+_1$, $+_2$ are defined by

$$\left(\alpha\ ;\ a\,\begin{smallmatrix} c \\ b \end{smallmatrix}\,d\right) +_1 \left(\beta\ ;\ e\,\begin{smallmatrix} b \\ f \end{smallmatrix}\,g\right) = \left(\beta+\alpha^g\ ;\ a+e\,\begin{smallmatrix} c \\ f \end{smallmatrix}\,d+g\right),$$

$$\left(\alpha\ ;\ a\,\begin{smallmatrix} c \\ b \end{smallmatrix}\,d\right) +_2 \left(\gamma\ ;\ d\,\begin{smallmatrix} u \\ w \end{smallmatrix}\,v\right) = \left(\alpha^w+\beta\ ;\ a\,\begin{smallmatrix} c+u \\ b+w \end{smallmatrix}\,v\right).$$

Calculation show that the interchange law for $D''_2$ is equivalent to the rule: (ii) $-\alpha+\beta+\alpha = \beta^{\delta\alpha}$, $\alpha, \beta \in A$. These rules (i) and (ii) are well known in the case $A, B$ are groups and determine $A$ as a *crossed* $B$*-module*. In the more general situation, we have the structure of a *crossed module over a groupoid*. The thin squares of $D''_2$ are the quintuples

$$\left(0\ ;\ a\,\begin{smallmatrix} c \\ b \end{smallmatrix}\,d\right).$$

There are clearly categories of crossed modules over groupoids and of double groupoids. We have indicated the existence of a functor from the first category to the second. *This functor is an equivalence of categories.* But crossed modules abound; two standard examples are a normal subgroup $A$ of $B$, with $B$ acting by conjugation on $A$, and the inner automorphism map $A \to \mathrm{Aut}\,A$. So we have also lots of examples of double groupoids.

The inverse equivalence $\gamma$ : (double groupoids) $\to$ (crossed modules over groupoids) is also easy to describe. If $G$ is a double groupoid, then we let $B$ be the groupoid $(G_1, G_0)$, and let  A  be the totally disconnected groupoid of elements  $\alpha$  in $G_2$  all of whose faces other than  $\partial^0 \alpha$  are identities, with composition given by  $\underset{2}{+}$ .  The face map  $\partial^0_1$  defines  $\delta : A \to B$ .

$$
\begin{array}{c} a \\ \boxed{\alpha} \end{array} \overset{\delta}{\longmapsto} a
$$

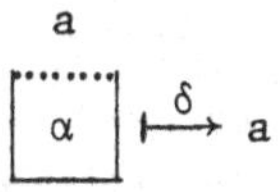

The operation of  B  on  A  is given by  $\alpha^b = \underset{2}{-} \epsilon_1 b \underset{2}{+} \alpha \underset{2}{+} \epsilon_1 b$ . The rules for a crossed module follow (cf.[12]), the crucial second rule being obtained by evaluating in two ways the composition

$$
\begin{array}{c}
\left[ \begin{array}{ccc} || & \beta & || \\ \underset{2}{-}\alpha & \Box & \alpha \end{array} \right] \begin{array}{c} \longrightarrow 2 \\ \downarrow \\ 1 \end{array}
\end{array}
$$

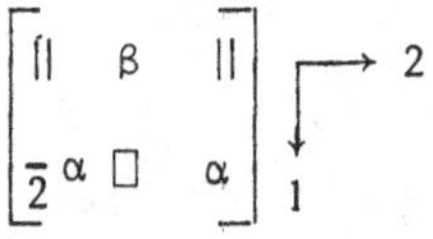

Now we find the non-trivial fact that  $\gamma$  applied to the double groupoid  $\rho\underset{\sim}{X}$  gives the well-known crossed module consisting of the relative homotopy groups  $\pi_2(X, X_1, x)$ , $x \in X_0$ , with the standard boundary  $\delta$  and operation of the fundamental groupoid.  Thus a pushout theorem for  $\rho\underset{\sim}{X}$  gives rise immediately to a pushout theorem for these crossed modules, and this is the main result of [6].

A special case of this result is that when  U , V , *and* W = U ∩ V *are path-connected and* (V , W) *is* 1 *-connected, then the excision map*  $\epsilon : \pi_2(V, W) \to \pi_2(X, U)$  *presents*  $\pi_2(X, U)$  *as the crossed*  $\pi_1 U$-*module induced from the crossed*  $\pi_1 W$-*module* $\pi_2(V, W)$  *by the map*  $\pi_1 W \to \pi_1 U$  *induced by inclusion* (cf [6] Theorem D).  We do not know any other method or proving this result.  Note that, like Whitehead's theorem which it implies, this is a non-abelian result.

It is difficult to see how to prove these results directly. One would somehow have to translate a subdivision of a square into the "linear" composition in  $\pi_2(X, U)$ :

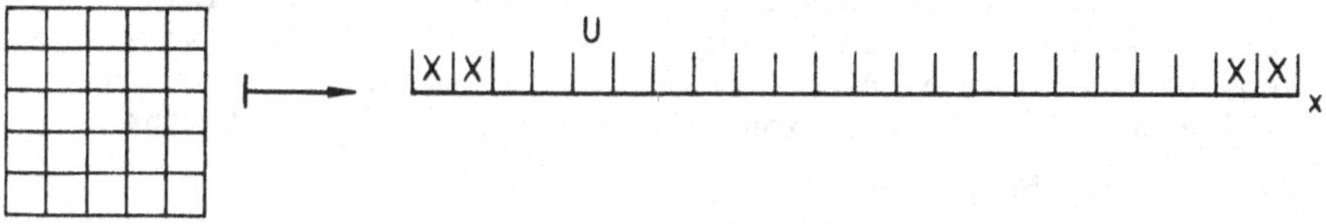

Essentially this job is done by the equivalence between double groupoids and crossed modules.

## 7. Still higher dimensions

The form of generalisation of the previous results to all dimensions is not hard to guess, but the proofs present a number of technical complications and require some new ideas. In particular, the above form of the homotopy addition lemma is not easy to handle even in the next dimension, since it would require subdivision of a three dimensional cube into 27 sub-cubes!

We start with the generalisation of the homotopy double groupoid. Let $\underset{\sim}{X}$ be a filtered space

$$\underset{\sim}{X} : X_0 \subset X_1 \subset X_2 \subset \ldots \subset X .$$

The cubical singular complex $KX$ of $X$ then has a subcomplex $R\underset{\sim}{X}$ which in dimension $n$ consists of the maps $\underset{\sim}{I}^n \to \underset{\sim}{X}$ of filtered spaces, where $I^n$ has its usual filtration by skeletons. By passing to filter homotopy classes (i.e. using homotopies through filtered maps) relative to the vertices of $I^n$ we obtain a quotient complex $\rho\underset{\sim}{X}$ with quotient map $p : R\underset{\sim}{X} \to \rho\underset{\sim}{X}$ . (The definitions of $\rho\underset{\sim}{X}$ in [9] is slightly different, but the two definitions coincide if $\pi_0 X_0 = X_0$ which is the useful case. The proofs in [9] are easily modified to the present definition.)

The complexes $R\underset{\sim}{X}$ and $\rho\underset{\sim}{X}$ have the usual face maps $\partial_i^\alpha (\alpha = 0 , 1 ; i = 1 , \ldots , n)$ and degeneracy maps $\varepsilon_j (j = 1,\ldots,n)$ of a cubical complex. Also $KX$ has "connections"

$$\Gamma_j : K_n X \to K_{n+1} X \qquad j = 1 , \ldots , n$$

induced by the maps

$$\Gamma'_j : I^{n+1} \to I^n$$

$$(t_1,\ldots,t_{n+1}) \mapsto (t_1,\ldots,t_{j-1} , \max\{t_j , t_{j+1}\} , t_{j+2},\ldots,t_{n+1})$$

and these connections are inherited by $R\underset{\sim}{X}$ and by $\rho\underset{\sim}{X}$ . This extra structure on the cubical complex is essential for our work, but seems to have been used previously only in [17]. The reason for calling the $\Gamma$'s "connections" is given in [13]. Note that $\Gamma'_1 : I^2 \to I^1$ can be pictured as

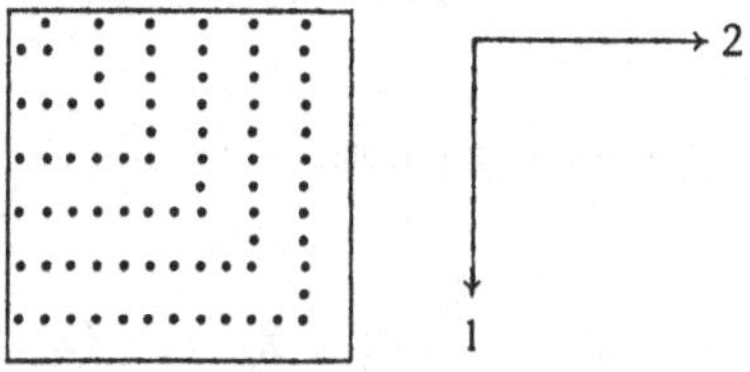

where the dotted lines denote lines of constancy.

The set $K_nX$ also has $n$ partial compositions $+_i$ with $a +_i b$ defined if and only if $\partial_i^1 a = \partial_i^0 b$ , $i = 1, \ldots, n$ . These compositions are clearly inherited by $R\underset{\sim}{X}$ . Further $\rho_1\underset{\sim}{X}$ inherits the operation $+_1$ , and in fact $\rho_1\underset{\sim}{X} = \pi_1(X_1, X_0)$ . . We have the following remarkable fact.

THEOREM 2 [9] *The operations* $+_i$ *on* $R_n\underset{\sim}{X}$ *induce groupoid structures* $+_i$ *on* $\rho_n\underset{\sim}{X}$ *such that any distinct* $+_i , +_j$ *satisfy the interchange law.*

The method of proof is to use collapsing arguments in $I^m$ to construct the required filtered homotopies.

There is an even more remarkable relation between $R\underset{\sim}{X}$ and $\rho\underset{\sim}{X}$ .

THEOREM 3 [9] *The quotient map* $p : R\underset{\sim}{X} \to \rho\underset{\sim}{X}$ *is a fibration in the sense of Kan.*

This "fibration theorem" is an important technical tool in the theory.

An object $G$ with the same kind of algebraic structure as $\rho\underset{\sim}{X}$ is called an *ω-groupoid*. Thus $G$ has all the operations $\partial_i^\alpha , \varepsilon_j , \Gamma_j , +_i$ satisfying what seems to be a formidable set of laws [8], but each of which has obvious geometric meaning. In a sense $\rho\underset{\sim}{X}$ is the *only* example of an ω-groupoid, in that every ω-groupoid is isomorphic to $\rho\underset{\sim}{X}$ for some $\underset{\sim}{X}$ . However, the proof of this fact (which generalises the construction of Eilenberg-MacLane complexes) is roundabout [9].

Clearly there is a category $\underset{\times}{G}$ of ω-groupoids, where a morphism $f : G \to H$ is a family of maps $f_n : G_n \to H_n$ , $n \geq 0$ , preserving all the structure.

We say the filtered space $\underset{\sim}{X}$ is *homotopy full* if the following conditions $(\phi_m)$ hold for all $m \geq 0$.

$(\phi_0)$ : The set $X_0$ meets each path-component of $X_r$, for all $r > 0$.

$(\phi_m)$ $(m \geq 1)$ : For all $r > m$ and $x_0 \in X_0$, $\pi_m(X_r, X_m, x_0) = 0$.

We can now state the generalisation to all dimensions of the Seifert-van Kampen theorem.

THEOREM 4 [9] *Let* $\underset{\sim}{X}$ *be a filtered space, and let* $U, V$ *be subsets of* $X$ *whose interiors cover* $X$. *Let* $W = U \cap V$ *and suppose that the filtered spaces* $\underset{\sim}{U}, \underset{\sim}{V}, \underset{\sim}{W}$, *induced by intersection from* $\underset{\sim}{X}$, *are homotopy full. Then the diagram of morphisms induced by inclusion*

$$\begin{array}{ccc} \rho\underset{\sim}{W} & \longrightarrow & \rho\underset{\sim}{U} \\ \downarrow & & \downarrow \\ \rho\underset{\sim}{V} & \longrightarrow & \rho\underset{\sim}{X} \end{array}$$

*is a pushout of* $\omega$*-groupoids.*

The proof follows a similar pattern to that of both Theorem 1 and the generalisation suggested in sections 4 and 5, except that the homotopy addition lemma has to be formulated and applied in a new way. The key concept is that of *thin* element in an $\omega$-groupoid, which is defined as an element which is a composite of elements of one of the forms $\varepsilon_i y$ or a repeated negative of some $\Gamma_j z$. This of course is a purely *algebraic* definition.

In $\rho_n\underset{\sim}{X}$ there is a *geometric* notion of thin element, namely one which has a representative $f : \underset{\sim}{I}^n \to \underset{\sim}{X}$ such that $f(I^n) \subset X_{n-1}$. It is important that *these two notions of thin element coincide* in $\rho\underset{\sim}{X}$. The proof uses Theorem 3 and most of the algebra of $\omega$-groupoids given in [8].

The cancellation argument discussed at the end of §4 is replaced by the easy fact that any composite of thin elements is thin, and the following result: *if* $k \in G_n$, *where* $G$ *is an* $\omega$*-groupoid, and* $k$ *and* $dk$ *are thin for any face operator* $d$ *not involving* $\partial^0_{n+1}, \partial^1_{n+1}$, *then* $k = \varepsilon_{n+1}\partial^0_{n+1}k$, *and hence* $\partial^0_{n+1}k = \partial^1_{n+1}k$. The proof of this also seems to require a full analysis of the algebra of $\omega$-groupoids.

## 8. Crossed complexes

Associated with a filtered space $\underset{\sim}{X}$ is the fundamental groupoid $C_1 = \pi_1(X_1, X_0)$ and the family of relative homotopy groups $C_n(x_0) = \pi_n(X_n, X_{n-1}, x_0)$ , $n \geq 2$ , $x_0 \in X_0$ . Further $C_1$ operates on the family $C_n$, $n \geq 2$ , and there are boundary maps $\delta : C_n \to C_{n-1}$ , $n \geq 2$ , such that $\delta^2 = 0$ . The part $\delta : C_2 \to C_1$ is a crossed module (over a groupoid). Also $C_n(x_0)$ is abelian and has trivial action by $\delta(C_2)$ for $n \geq 3$ . Such a gadget we call a *crossed complex*. That associated as above with a filtered space $\underset{\sim}{X}$ we call the *homotopy crossed complex* of $\underset{\sim}{X}$ and write it $\pi\underset{\sim}{X}$ .

Such structures have been considered by Blakers [1] under the name 'group system' and by Whitehead [27] under the name 'homotopy system'. Their utility in interpreting the cohomology of groups has been shown recently by Huebschmann [22], Holt [21], and indeed they were in effect considered also by Gerstenhaber [18] for the same reason.

The main algebraic result on $\omega$-groupoids, and the main result of [8] is:

THEOREM 5 *There is an equivalence of categories*

$$\gamma : \omega\text{-}groupoids \longrightarrow crossed\ complexes .$$

This theorem leads to an interpretation of $\rho\underset{\sim}{X}$ , as follows.

THEOREM 6 [9] *If* $\underset{\sim}{X}$ *is any filtered space, and* $\gamma$ *is as in Theorem 5, then there is a natural isomorphism of crossed complexes*

$$\gamma\rho\underset{\sim}{X} \cong \pi\underset{\sim}{X} .$$

Thus Theorem 4 leads to a pushout theorem for the homotopy crossed complex $\pi\underset{\sim}{X}$ , and it is very special cases of this theorem which yield the results given in section 1.

These theorems suggest the importance of crossed complexes in algebraic topology. It is therefore of interest that there are four algebraic categories known to be equivalent to crossed complexes, namely $\omega$-groupoids as mentioned above, $\infty$-groupoids [11], cubical T-complexes [10], and simplicial T-complexes [1]. All these equivalences are non-trivial. In a cubical (or simplicial) T-complex $(K, T)$ we are given subsets

$$T_n \subset K_n , \quad n \geq 1 ,$$

of elements called *thin* and satisfying Keith Dakin's axioms [15]

T1) degenerate elements are thin,

T2) every box has a unique thin filler,

T3) if all faces but one of a thin element are thin, so also is the last face.

It is surprising how much algebra is generated by these deceptively simple axioms. (D.W. Jones has defined a category of polyhedral T-complexes and shown that these also are equivalent to simplicial T-complexes.)

It should be stressed that the homotopy full conditions in Theorem 4 limit its applications. In fact we have recently found that a filtered space $\underset{\sim}{X}$ is homotopy full if and only if the induced map $\pi_0 X_m \to \pi_0 X_r$ is surjective for $m = 0$ and bijective for $r > m \geq 1$, and the pairs $(X_r, X_m)$ based at any $x_0 \in X_0$ are m-connected for $r > m \geq 1$. It is thus reasonable to regard the above theorems as giving algebraicisations, generalisations and strengthening of aspects of homotopy theory up to and around the homotopy addition lemma and relative Hurewicz theorem. In particular, there seems no hope of determining $\pi_3(S^2)$ as a consequence of Theorem 4.

One interesting aspect of Theorem 4 is that it links results in various dimensions, and some of them non-abelian, in the form of a colimit theorem rather than an exact sequence or spectral sequence. The advantage of a colimit theorem is that it gives a complete answer, rather than an answer up to extension, and the prototype of such a result is Theorem 1 which computes $\pi_1(U \cup V, x_0)$ precisely even when $U \cap V$ is not path-connected. Theorem 4 is obtained by an algebraic gadget $\rho\underset{\sim}{X}$ which incorporates information in all dimensions. It would be interesting to know if this technique is of wider applicability.

## REFERENCES


1. N. ASHLEY, *T-complexes and crossed complexes*, Ph.D. Thesis, University of Wales, (1978).

2. A.L. BLAKERS, 'Some relations between homology and homotopy groups', *Ann. of Math.* 49 (1948), 428-461.

3. R. BROWN, 'Groupoids and van Kampen's theorem', *Proc. London Math. Soc.* (3) 17 (1967), 385-40.

4. R. BROWN, *Elements of Modern Topology*, McGraw Hill (Maidenhead) (1968).

5. R. BROWN, 'On the second relative homotopy group of an adjunction space: an exposition of a theorem of J.H.C. Whitehead', *J. London Math. Soc.* (2) 22 (1980) 146-152.

6. R. BROWN and P.J. HIGGINS, 'On the connection between the second relative homotopy groups of some related spaces', *Proc. London Math. Soc.* (3) 36 (1978), 193-212.

7. R. BROWN and P.J. HIGGINS, 'Sur les complexes croisés, ω-groupoids, et T-complexes', *C.R. Acad. Sc. Paris* 285 (1977) 997-999; 'Sur les complexes croisés d'homotopie associés à quelques espaces filtrés, *ibid* 286 (1978) 91-93.

8. R. BROWN and P.J. HIGGINS, 'The algebra of cubes', *J. Pure and Applied Alg.* 21 (1981) 233-260.

9. R. BROWN and P.J. HIGGINS, 'Colimits of relative homotopy groups', *J. Pure and Applied Algebra* 22 (1981) 11-41.

10. R. BROWN and P.J. HIGGINS, The equivalence of ω-groupoids and cubical T-complexes', *Cah. Top. Géom. Diff.* 22 (1981), 349-370.

11. R. BROWN and P.J. HIGGINS, 'The equivalence of ∞-groupoids and crossed complexes', *Cah. Top. Géom. Diff.* 22 (1981), 371-386.

12. R. BROWN and C.B. SPENCER, $\mathcal{G}$-groupoids, crossed modules and the fundamental groupoid of a topological group', *Proc. Kon. Ned. Akad. v. Wet.* 79 (1976) 296-302.

13. R. BROWN and C.B. SPENCER, 'Double groupoids and crossed modules', *Cah. Top. Géom. Diff.* 17 (1976) 343-362.

14. R.H. CROWELL, 'On the van Kampen theorem', *Pacific J. Math.* 9 (1959) 43-50.

15. K. DAKIN, *Kan complexes and multiple groupoid structures*, Ph.D. Thesis, University of Wales (1977).

16. C.H. EHRESMANN, *Catégories et structures*, Dunod (Paris) (1965).

17. M. ÉVERARD, 'Homotopie des complexes simpliciaux et cubiques', preprint, (1972).

18. M. GERSTENHABER, 'A uniform cohomology for algebra', *Proc. Nat. Acad. Sci.* 51 (1964) 626-629.

19. P.R. HEATH, *An introduction to homotopy theory via groupoids and universal constructions*, Queen's papers in pure and applied mathematics No. 49, Queen's University (1978).

20. P.J. HIGGINS, *Categories and groupoids*, Van Nostrand (1971).

21. D.F. HOLT, 'An interpretation of the cohomology groups $H^n(G, M)$', *J. Algebra* 60 (1979) 307-318.

22. J. HUEBSCHMANN, 'Crossed n-fold extensions of groups and cohomology', *Comm. Math. Helv.* 55 (1980) 302-314.

23. A. RAZAK, *Union theorems for double groupoids and groupoids; some generalisations and applications*, Ph.D. Thesis, University of Wales (1976).

24. C.B. SPENCER, 'An abstract setting for homotopy pushouts and pullbacks', *Cah. Top. Géom. Diff.* 18 (1977) 409-430.

25. J.H.C. WHITEHEAD, 'On adding relations to homotopy groups', *Ann. of Math.* 42 (1941) 409-428.

26. J.H.C. WHITEHEAD, 'Note on a previous paper entitled "On Adding relations to homotopy groups"', *Ann. of Math.* 47 (1946) 806-810.

27. J.H.C. WHITEHEAD, 'Combinatorial homotopy II', *Bull. American Math. Soc.* 55 (1949) 453-496.

28. J.H.C. WHITEHEAD, 'A certain exact sequence', *Ann. of Math.* 52 (1950) 51-110.

# Part 4: 4-manifolds

# Actions of compact connected groups on 4-manifolds

P. ORLIK

## 1. Introduction

The best known examples of manifolds have large symmetry groups. In the case of 3-manifolds, $S^3$, $P^3$, $S^2 \times S^1$, $T^3$ and the lens spaces all admit actions by relatively "large" compact Lie groups. The manifolds of Seifert [12] which are now known to be among the building blocks of all 3-manifolds may be viewed as 3-manifolds with (local) circle action. The purpose of this talk is to present a similarly tractable class of 4-manifolds. We shall outline the equivariant classification of 4-manifolds with compact connected Lie group actions and indicate the results and remaining questions about their topological classification. For the definitions and the basic terminology of transformation groups see Bredon's book [1]. For simplicity we shall assume that all manifolds are closed, oriented and smooth, and all actions are differentiable. A theorem of Eisenhardt [2, p. 239] asserts that if $M^m$ admits a G-action, then $\dim G \leq \frac{1}{2}m(m+1)$ and if equality holds, then $M^m$ is $S^m$ (or $P^m$). In case $m = 4$ we have $\dim G \leq 10$. The classification theorem of compact connected Lie groups [1, p. 30] asserts that $G \simeq (T^k \times G')/N$ where $G'$ is a product of simply connected simple Lie groups and $N$ is a finite central subgroup. If we allow our actions to be almost effective, i.e. allow a finite subgroup to fix every point, then we need only consider groups of the form $T^k \times G'$. There is an additional consideration restricting the possible groups. Let $m(G)$ be the smallest dimension of a manifold with G-action. This dimension was computed by Mann [4], and it is sufficient to consider groups with $m(G) \leq 4$. The following is a list of groups satisfying both conditions:

| G: | Spin(5) | SU(3) | $S^1\times$Spin(4) | Spin(4) | $T^2\times$Spin(3) | $T^4$ | $S^1\times$Spin(3) | $T^3$ | Spin(3) | $T^2$ | $S^1$ |
|---|---|---|---|---|---|---|---|---|---|---|---|
| dimG: | 10 | 8 | 7 | 6 | 5 | 4 | 4 | 3 | 3 | 2 | 1 |
| m(G): | 4 | 4 | 4 | 3 | 4 | 4 | 3 | 3 | 2 | 2 | 1 |

## 2. Equivariant classification.

The dimension of a principal orbit is an invariant, so we may organise the presentation accordingly. If $(M,G)$ has a 4-dimensional orbit, then it is a homogeneous space listed below, where $G$ is the Lie group, $H$ the closed subgroup and $M \simeq G/H$ the homogeneous space.

| | | | | | | |
|---|---|---|---|---|---|---|
| G: | Spin(5) | SU(3) | $S^1 \times$Spin(4) | Spin(4) | $T^4$ | $S^1 \times$Spin(3) |
| H: | Spin(4) | U(2) | Spin(3) | $T^2$ | 1 | N |
| M: | $S^4$ | $CP^2$ | $S^1 \times S^3$ | $T^2 \times S^2$ or $S^1 \times K^3$ | $T^4$ | $S^1 \times K^3$ |

Here $N$ is a finite subgroup and $K^3$ is a finite quotient of $S^3$.

If $(M,G)$ has 3-dimensional principal orbits, then the orbit space $M^*$ is a 1-manifold. If $M^* = S^1$, then all orbits are principal with isotropy type $H$. Thus $M$ is a bundle over $S^1$ with fiber the 3-manifold $G/H$ and structure group $N(H)/H$, where $N(H)$ is the normalizer of $H$, see [1, p.206]. If $M^*$ is a closed interval, then in addition to the principal isotropy type $H$ we have the isotropy types $U_0$ and $U_1$ of the orbits corresponding to the end points. We may assume $H \subset U_i$. Then $U_i/H$ is an $r_i$ sphere for some $r_i$, $i = 0, 1$ and $M$ is obtained by the equivariant identification of the common boundary $G/H$ of $G \times_{U_0} D^{r_0+1}$ and $G \times_{U_1} D^{r_1+1}$. The resulting manifolds are classified equivariantly by the set of distinct identification maps. The latter is in one-to-one correspondence with the double coset space $N_0\backslash N(H)/N_1$, where $N_i = N(H) \cap N(U_i)$. This gives the equivariant classification in principle but the details, currently studied by Parker [11], require the consideration of a large number of cases.

If $M$ has 2-dimensional principal orbits, then $G = T^2$ or Spin(3). The equivariant classification of 4-manifolds with $T^2$-action was obtained in [6] in terms of the weighted orbit space and certain other numerical invariants. If $G = \text{Spin}(3)$, or what amounts to the same, $G = SO(3)$ then the principal isotropy type is $O(2)$ or $SO(2)$. The first possibility is ruled out, since local considerations do not allow fixed points, so $M$ is a $P^2$-bundle over $M^*$ with trivial structure group and thus non-orientable. The principal isotropy type is therefore $SO(2)$, with orbit $S^2$. In addition there may be fixed points and $P^2$-orbits with isotropy $O(2)$. The orbit space is a 2-manifold with boundary. Each component of the boundary consists entirely of fixed points or entirely of the images of $P^2$-orbits. Let $f \geq 0$ be the number of the boundary components of $M^*$ with fixed points, let $p \geq 0$ be the number of boundary components of $M^*$ with $P^2$-orbits and let $g$ be the genus of $M^*$. If $f + p = 0$, then $M$ is an $S^2$ bundle over the closed 2-manifold $M^*$ with structure group $N(SO(2))/SO(2) = \mathbb{Z}_2$. If $M^*$ is orientable then $M \simeq M^* \times S^2$. If $M^*$ is non-orientable then $M \simeq M^* \tilde{\times} S^2$, the $S^2$ bundle over $M^*$ with orientable total space. If $f + p > 0$ then $(g, \varepsilon, f, p)$ is a complete set of equivariant invariants for the action, where $\varepsilon = 0$ or $n$ according to the orientability of $M^*$.

If $(M,G)$ has 1-dimensional principal orbits, then $G = S^1$. The equivariant classification of 4-manifolds with $S^1$-action was obtained by Fintushel [3] in terms of the weighted orbit space and certain other numerical invariants.

## 3. Topological classification.

The dimension of a principal orbit is not a topological invariant. Nevertheless it is convenient to determine first the topological equivalences among the equivariant equivalence classes of the action of one fixed group. We have already identified the 4-dimensional homogeneous spaces. The topological classification of actions with 3-dimensional orbits is studied by Parker [11].

Next suppose $G = SO(3)$. There are several possibilities. First assume that $M^*$ is oriented. If $f + p = 0$ then $M \simeq M^* \times S^2$. If $f > 0$, then $M$ is an equivariant connected sum:

$$M \simeq S^4 \# (S^3 \times S^1)_{f+g-1} \# (P^3 \times S^1)_p \ .$$

Here the $S^4$ summand is needed only if $g = p = 0$ and $f = 1$. To see this we identify certain elementary manifolds. Suppose $M^* = D^2$ with $f = 1$, $g = p = 0$. Then $M \simeq S^4$. The action on $S^4$ is given by representing $S^4$ as a join of $S^2$ and $S^1$ and letting $G$ act on the $S^2$ factor. Next suppose $M^* = S^1 \times I$ with $f = 2$, $g = p = 0$. Then $M \simeq S^3 \times S^1$ with $G$ acting on the $S^3$ factor. This action of $G$ on $S^3$ has two fixed points and all other orbits are principal. If $M^* = S^1 \times I$ with $f = 1$, $p = 1$, $g = 0$, then $M \simeq P^3 \times S^1$ by a similar argument. Finally, if $M^*$ is a torus with one boundary component of fixed points, $f = 1$, $g = 1$, $p = 0$, then $M \simeq S^3 \times S^1 \# S^3 \times S^1$. The dotted arc $\Sigma^*$ is the image of a 3-sphere $\Sigma$. Cutting $M$ open along $\Sigma$ and filling in the two boundary components equivariantly with 4-balls results in a G-manifold $N$ with orbit space $N^* = S^1 \times I$, $f = 2$. Thus $N \simeq S^3 \times S^1$ and $M$ is obtained from $N$ by removing two balls and identifying the boundaries, so $M \simeq N \# S^3 \times S^1 \simeq S^3 \times S^1 \# S^3 \times S^1$.

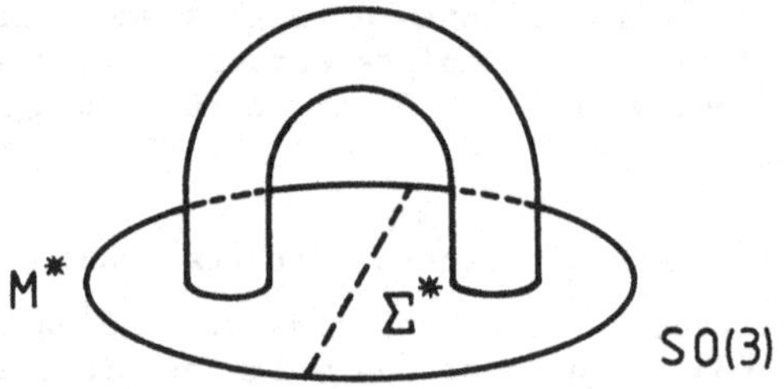


It remains to consider the case $f = 0$, $p > 0$. Let $M^* = D^2$. Let $K^*$ be an interior circle. Then $K \simeq S^1 \times S^2$, $M_1 \simeq D^2 \times S^2$ and $M_2 \simeq Q \times S^1$ where $Q = P^3 - \mathring{D}^3$ is a projective 3-space with an open ball removed. Thus $M \simeq M_1 \underset{K}{\cup} M_2$ where the identification maps $\partial D^2$ onto $S^1$ and $S^2$ onto $\partial Q$ equivariantly.

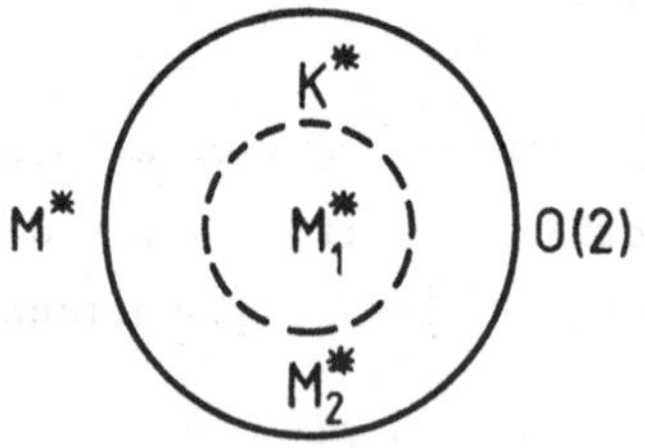


This shows for example that $\pi_1 M \simeq \mathbb{Z}_2$, and the integral homology groups are $H_0 M \simeq \mathbb{Z}$, $H_1 M \simeq \mathbb{Z}_2$, $H_2 M \simeq \mathbb{Z}_2$, $H_3 M \simeq 0$, $H_4 M \simeq \mathbb{Z}$. In fact $M$ is obtained from $P^3 \times S^1$ by a surgery on the second factor. If $M^* = I \times S^1$, $p = 2$, $f = 0$, then $M \simeq (P^3 \# P^3) \times S^1$ since $M$ is the union of two copies of $Q \times S^1$ along the common boundary, $S^2 \times S^1$. In general if $f = 0$, $p > 0$, then $M$ is obtained from an equivariant connected sum by replacing $D^3 \times S^1$

by $Q \times S^1$. This corresponds to looking at the manifold with $f = 1$ and $p - 1$ components of $P^2$-orbits, removing a tubular neighbourhood of the fixed point set and replacing it by $P^2$-orbits, i.e. by $Q \times S^1$.

If $M^*$ is non-orientable, then the $S^2$ bundle over each orientation reversing path is twisted. The total space $N$ of the twisted $S^2$-bundle over a Moebius band is the mapping cylinder of the 2-fold cover of the non-orientable 3-handle by $S^1 \times S^2$ with $\partial N = S^1 \times S^2$. Now $M^* \approx R^2 \# kP^2$ is the connected sum of an orientable surface $R^2$ with $k = 1$ or $2$ copies of $P^2$. Thus we may obtain $M$ by first constructing the manifold $M'$ over $R$ with the given values of $f$ and $p$, and then performing $k$ times the operation which removes $D^2 \times S^2$ from $M'$ and replaces it by a copy of $N$.

For $G = T^2$ the topological classification is more difficult. Let $F$ denote the fixed point set, $E$ the collection of orbits with finite non-trivial isotropy group and $C$ the collection of orbits with isotropy group isomorphic to the circle. It was shown in [7] that in case $F \neq \phi$, the manifold is the equivariant connected sum of building blocks. Some of these elementary $T^2$-manifolds were later identified by Pao [8] while others are still not completely understood. The case $C \cup F = \phi$, i.e. where only finite non-trivial orbits occur, resembles 3-dimensional Seifert manifolds and may be classified along the same lines [7]. The remaining case, $F = \phi$, $C \neq \phi$ has proved to be the least tractable. Pao [9] has made some progress and more recently Melvin [5] has obtained some additional interesting results, but there are still many unanswered questions.

One significant contribution has been made in the topological classification of 4-manifolds with $S^1$-action. Let $M$ be a homotopy 4-sphere with an effective smooth $S^1$-action. There are at most four orbit types: fixed points $(F)$, principal orbits and possibly exceptional orbits with isotropy $\mathbb{Z}_n$ (called $E_n$) and with isotropy $\mathbb{Z}_m$ (called $E_m$). Either $F = S^2$ or $F = S^0$. If $F = S^2$ then the orbit space $M^*$ is a homotopy 3-disk and $E_n \cup E_m = \phi$. If $F = S^0$ then $M^*$ is a homotopy 3-sphere. If there is only one exceptional orbit type, then $E_n^* \cup F^*$ is an arc in $M^*$. If there are two exceptional orbit types, then $m$ and $n$ must be relatively prime and $E_n^* \cup F^* \cup E_m^*$ is a circle in $M^*$.

Conversely, any orbit structure described above corresponds to a smooth $S^1$-action on a homotopy 4-sphere.

Pao [10] shows that if $M^* = D^3$ or $M^* = S^3$, then $M = S^4$. In particular the knot type $(M^*, E_n^* \cup F^* \cup E_m^*)$ does not matter. Since the image of the exceptional set is unknotted for orthogonal actions, this shows that there are non-orthogonal $S^1$-actions on $S^4$. The main tool is a surgery move on M which results in a new $S^1$-manifold M' such that M and M' are diffeomorphic, but the equivariant data for M' are different. Repeated application shows that M is diffeomorphic to an $S^1$-manifold with no exceptional orbits. The latter is clearly $S^4$.

## REFERENCES

1. G. BREDON, *Introduction to compact transformation groups*, Academic Press, New York, London 1972.

2. L.P. EISENHART, *Riemannian geometry*, Princeton University Press, 1926.

3. R. FINTUSHEL, 'Classification of circle actions on 4-manifolds', *Trans. Amer. Math. Soc.* 242 (1978), 377-390.

4. L. MANN, 'Gaps in the dimensions of transformation groups', *Illinois J. Math.* 19 (1966), 532-546.

5. P. MELVIN, 'On 4-manifolds with singular torus actions', preprint.

6. P. ORLIK and F. RAYMOND, 'Actions of the torus on 4-manifolds I', *Trans. Amer. Math. Soc.* 152 (1970), 531-559.

7. P. ORLIK and F. RAYMOND, 'Actions of the torus on 4-manifolds II', *Topology* 13 (1974), 89-112.

8. P.S. PAO, 'The topological structure of 4-manifolds with effective Torus actions I', *Trans. Amer. Math. Soc.* 227 (1977), 279-317.

9. P.S. PAO, 'The topological structure of 4-manifolds with effective torus actions II', *Illinois J. Math.* 21 (1977), 883-894.

10. P.S. PAO, 'Non-linear circle actions on the 4-sphere and twisting spun knots', *Topology* 17 (1978), 291-296.

11. J. PARKER, '4-dimensional G-manifolds with 3-dimensional orbits', Thesis, University of Wisconsin, 1980.

12. H. SEIFERT, 'Topologie dreidimensionaler gefaserter Räume', *Acta math.* 60 (1933), 147-238.